Illuminate Publishing

D0529189

WJEC
AS Biology

Study and Revision Guide

Gareth Rowlands

Published in 2011 by Illuminate Publishing Ltd, P.O Box 1160, Cheltenham, Gloucestershire GL50 9RW

Orders: Please visit www.illuminatepublishing.com
or email sales@illuminatepublishing.com

British Library Cataloguing in Publication Data

A catalogue record for this book is available from the British Library

ISBN 978-0-9568401-4-1

Printed by T J International, Padstow, Cornwall

The publisher's policy is to use papers that are natural, renewable and recyclable products made from wood grown in sustainable forests. The logging and manufacturing processes are expected to conform to the environmental regulations of the country of origin.

Every effort has been made to contact copyright holders of material reproduced in this book. If notified, the publishers will be pleased to rectify any errors or omissions at the earliest opportunity.

This material has been endorsed by WJEC and offers high quality support for the delivery of WJEC qualifications. While this material has been through a WJEC quality assurance process, all responsibility for the content remains with the publisher.

Editor: Geoff Tuttle

Design and layout: Nigel Harriss

Permissions
cover: ©Fotolia; TheSupe87: p47; ©Shutterstock: Yuliyan Velchev; Roman Sotola; Denis Cristo; Bakelyt

Acknowledgements
I am very grateful to the team at Illuminate Publishing for their professionalism, support and guidance throughout this project. It has been a pleasure to work so closely with them.

The author and publisher wish to thank:

Dr John Ford for his thorough review of the book and expert insights and observations.

We are indebted to Mike Ebbsworth of WJEC whose unstinting help and encouragement from the beginning made this whole undertaking possible.

Dr Janet Jones of WJEC.

Contents

How to use this book

As a Principal Examiner for the WJEC specification I have written this study guide to help you be aware of what is required, and structured the content to guide you through to success in the WJEC Biology AS level examination.

Knowledge and Understanding

The **first section** of the book covers the key knowledge required for the examination.

There are notes for the compulsory sections of each examination:

> BY1 – Basic Biochemistry and Cell Structure
>
> BY2 – Biodiversity and Physiology of Body Systems.

In addition, I have tried to give you additional pointers so that you can develop your work:

- Any of the terms in the WJEC specification can be used as the basis of a question, so I have highlighted those terms and offered definitions.
- There are 'Quickfire' questions designed to test your knowledge and understanding of the material.
- I have offered examination advice based on experience of what candidates need to do to attain the highest grades.

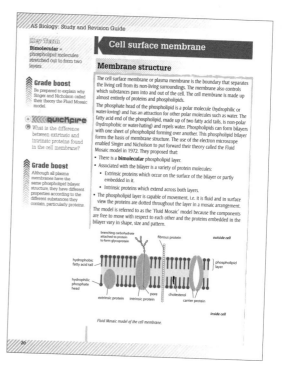

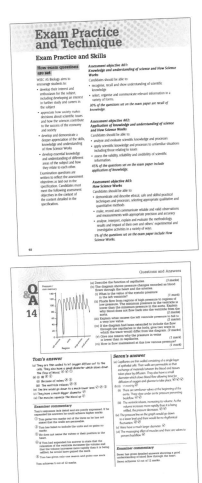

Exam Practice and Technique

The **second section** of the book covers the key skills for examination success and offers you examples based on real-life responses to examination questions. First you will be guided into an understanding of how the examination system works, and then offered clues to success.

A variety of structured and essay questions are provided in this section. Each essay includes the marking points expected followed by actual samples of candidates' responses. A variety of structured questions are also provided, together with typical responses and comments. They offer a guide as to the standard that is required, and the commentary will explain why the responses gained the marks that they did.

Most importantly, I advise that you should take responsibility for your own learning and not rely on your teachers to give you notes or tell you how to gain the grades that you require. You should look for additional notes to support your study into WJEC Biology.

I advise that you look at the WJEC website www.wjec.co.uk. In particular, you need to be aware of the specification. Look for specimen examination papers and mark schemes. You may find past papers useful as well.

Good luck with your revision.

Gareth Rowlands

Knowledge and Understanding

BY1 Basic Biochemistry and Cell Structure

The BY1 unit incorporates biochemistry and cell structure which is fundamental to the functioning of living organisms. The function of molecules depends on their properties. A molecule gets its properties from its structure. A knowledge of basic cell structure is essential in the understanding of the transport mechanisms involved in the exchange of molecules between the cell and its surroundings. In cells metabolic reactions take place rapidly involving thousands of simultaneous reactions. Order and control is essential if reactions are prevented from interfering with each other. These features of metabolism are made possible by the action of enzymes. Chromosomes are made up mainly of DNA. Genetic information needs to be copied and passed on to the daughter cells by the process of cell division known as mitosis.

Revision checklist

Tick column 1 when you have completed brief revision notes.
Tick column 2 when you think you have a good grasp of the topic.
Tick column 3 during final revision when you feel you have mastery of the topic.

		1	2	3	Notes
Enzymes					
p26	Enzyme structure				
p27	Factors affecting the rate of enzyme action				
p29	Enzyme inhibition				
p30	Medical and industrial applications of enzymes				
Nucleic acids					
p32	The structure of DNA				
p33	The structure of RNA				
Cell division					
p34	Mitosis				
p36	Meiosis				
p36	Comparison of mitosis and meiosis				

Grade boost

You should be able to recognise and understand structural formulae but not reproduce them. However, be prepared to use structural formulae if they are provided in an exam question.

quickfire

① With reference to carbohydrates explain the difference between a condensation reaction and hydrolysis.

α-glucose

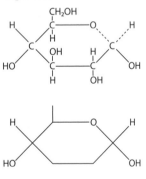

(simplified)

Molecular structure of α glucose.

Biochemistry

Carbohydrates

Carbohydrates are organic compounds containing the elements carbon, hydrogen and oxygen. Many organic molecules, including carbohydrates, are made up of a chain of individual molecules each of which is given the general name monomer. Polymers are longer chains of repeating monomer units. In carbohydrates the basic monomer unit is called a monosaccharide. Two monosaccharides combine to form a disaccharide.

Many monosaccharide molecules combine to form a polysaccharide.

Monosaccharides

Monosaccharides are relatively small organic molecules and provide the building blocks for the larger carbohydrates. Monosaccharides have the general formula $(CH_2O)n$ and their name is determined by the number of carbon atoms in the molecule (n). A triose sugar has three carbon atoms, a pentose sugar five carbon atoms. Glucose is a hexose sugar.

All sugars share the formula $C_6H_{12}O_6$ but they differ in their molecular structure. Monosaccharides usually exist as ring structures when dissolved in water. Glucose exists as two **isomers**, the α form and the β form. These different forms result in considerable biological differences when they form polymers such as starch and cellulose.

Disaccharides

Disaccharides consist of two monosaccharide units linked together with the formation of a glycosidic bond *and the elimination of water*. This is called a condensation reaction. Disaccharides can be formed by the joining of two similar monosaccharides or by the joining of two different monosaccharides.

Glucose joined to glucose forms maltose.

Glucose joined to fructose forms sucrose.

Glucose joined to galactose forms lactose.

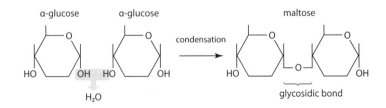

Formation of a glycosidic bond when two glucose molecules combine to form maltose.

Polysaccharides

Polysaccharides are an example of large, complex molecules called polymers. They are formed from very large numbers of monosaccharide units linked together by glycosidic bonds.

Starch is a storage polysaccharide found in plant cells in the form of starch grains. These are found in seeds and storage organs such as potato tubers. Starch is made up of many α glucose molecules held together. It is an ideal storage molecule because it is compact and can be stored in a small space; is insoluble and does not draw water towards it by osmosis.

Starch is made up of two polymers amylose and amylopectin. Amylose is linear (unbranched) and coils into a helix, whereas amylopectin is branched and fits inside the amylose.

The main storage product in animals is called glycogen, sometimes called animal starch and is very similar to amylopectin. Both starch and glycogen are readily hydrolysed to α glucose, which is soluble and can then be transported to areas where energy is needed.

Cellulose is a structural polysaccharide and is a major component of plant cell walls.

Cellulose consists of many long parallel chains of β glucose molecules cross-linked to each other by hydrogen bonds. Being made up of β glucose units, the chain has adjacent glucose molecules rotated by 180°. This allows hydrogen bonds to be formed between the hydroxyl groups of adjacent parallel chains and helps to give cellulose its structural stability. These chains are grouped together into microfibrils a number of which are arranged in parallel groups called fibres. The large number of hydrogen bonds present contribute to the strength and rigidity of plant cell walls.

Chitin is a polysaccharide found in insects. It is similar to cellulose but has amino acids added to form a mucopolysaccharide. It is strong, waterproof and lightweight and forms the exoskeleton of insects.

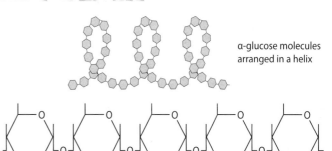

α-glucose molecules arranged in a helix

glycosidic bond

Structure of a molecule of starch.

>> **Pointer**

All monosaccharides and some disaccharides, such as maltose, are reducing sugars. Describe the Benedict's test for a reducing sugar. How does the test for a non-reducing sugar, such as sucrose, differ?

Grade boost

In β glucose units the positions of the –H group and the –OH group on a single carbon atom are reversed. In β glucose the –OH group is above, rather than below. This means that to form glycosidic links each β glucose molecule must be rotated by 180° compared to the one next to it.

quickfire

② State whether these carbohydrates are mono-, di-, or polysaccharides and give each their role in living organisms. State whether each occurs in plants, animals or both: lactose, cellulose, glucose, glycogen.

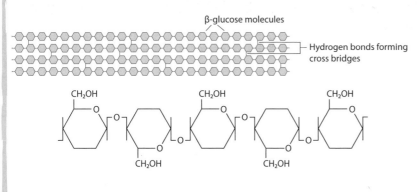

β-glucose molecules

Hydrogen bonds forming cross bridges

Structure of a molecule of cellulose.

③ Name the products formed and the type of bond that is broken when a triglyceride is broken down.

④ Suggest why parts of organisms that move, such as seeds, use lipids as an energy store rather than carbohydrates.

⑤ What is meant by metabolic water?

Lipids

Fats

Like carbohydrates, lipids also contain carbon, hydrogen and oxygen but in proportion to the carbon and hydrogen they contain less oxygen. They are **non-polar** compounds and so are insoluble in water.

Triglycerides are formed by condensation reactions between glycerol and fatty acids. A triglyceride consists of one molecule of glycerol and three fatty acid molecules. The glycerol molecule in a lipid is always the same but the fatty acid component varies. In this reaction, water is removed and an oxygen bond, known as an ester bond, is formed between the glycerol and fatty acid.

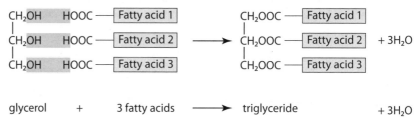

Formation of triglyceride.

The glycerol molecule is always the same but there are many different fatty acids that might react with glycerol. There are of two main kinds of fatty acids:

- Saturated fatty acids where all the carbon atoms are joined by a single bond.
- Unsaturated fatty acids contain one or more double bonds and therefore have fewer hydrogen atoms than they might.

A high intake of fat, notably saturated fats, is a contributory factor in heart disease.

Chemical properties

- They are insoluble in water but dissolve in organic solvents such as acetone and alcohols.
- Fats are solid at room temperature, whereas oils are liquids.

Functions

- Energy storage. Fats are an efficient energy store in both plants and animals. One gram of fat when oxidised yields approximately twice as much energy as the same mass of carbohydrate.
- Triglycerides also produce a lot of metabolic water when oxidised. This is important in desert animals such as the camel.
- Protection of delicate internal organs such as kidneys.
- Insulation. Fats are poor conductors of heat and when stored under the skin, they help retain body heat.
- Waterproofing. Fats are insoluble in water and are important in land organisms such as insects where the waxy cuticle cuts down water loss. Leaves also have a waxy cuticle to reduce transpiration.

Phospholipids

Phospholipids are important in the formation and functioning of membranes in cells. They are similar to triglycerides with one of the fatty acid groups replaced by a phosphate group.

- The lipid part is non-polar and insoluble in water (**hydrophobic**).
- The phosphate group is **polar** and dissolves in water (**hydrophilic**).

Phospholipids allow lipid-soluble substances to enter and leave a cell and prevent water-soluble substances entering and leaving the cell.

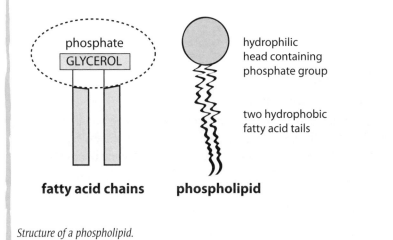

fatty acid chains   **phospholipid**

Structure of a phospholipid.

Key Terms

Hydrophilic = attracts water.

Hydrophobic = repels water.

quickfire

⑥ State two differences between a triglyceride and a phospholipid.

quickfire

⑦ Which end of the phospholipid molecule lies to the outside of the membrane?

quickfire

⑧ State through which part of the membrane each of the following passes in order to enter or leave a cell:
1. Sodium ion.
2. A lipid soluble molecule.

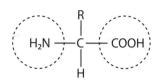

Structural formula of a generalised amino acid.

>> **Pointer**

Proteins carry out a range of biological activities and include enzymes, antibodies, hormones, carrier and transport proteins, as well as structural proteins.

>> **Pointer**

Make sure you are familiar with the different bonds involved in each of the different levels of protein structure.

>> **Pointer**

Each protein is different because of its 3D shape. This allows the protein to recognise and be recognised by other molecules, for example the combination of enzymes and substrates.

quickfire

(9) List the four bonds present in the tertiary structure of a protein.

Tertiary structure of protein.

Proteins

- Proteins differ from carbohydrates and lipids in that in addition to carbon, hydrogen and oxygen, they always contain nitrogen. Many proteins also contain sulphur and sometimes phosphorous.
- Proteins are large compounds built up of sub-units called amino acids. About 20 different amino acids are used to make up proteins. There are thousands of different proteins and their shape is determined by the specific sequence of amino acids in the chain.
- All amino acids have the same basic structure in that each possesses an amino group, NH_2, at one end of the molecule, and a carboxyl group, -COOH, at the other end. It is the R group which differs from one amino acid to another.

The peptide bond

Proteins are built up from a linear sequence of amino acids. The amino group of one amino acid reacts with the carboxyl group of another with the elimination of water. The bond that is formed is called a peptide bond and the resulting compound is a dipeptide. A number of amino acids joined in this way is called a polypeptide.

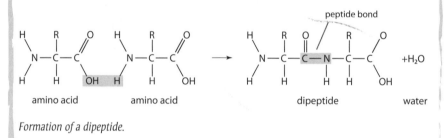

Formation of a dipeptide.

Protein structure

Four levels of protein structure exist:

1. The primary structure of a protein is the sequence of amino acid in its polypeptide chain. The proteins differ from each other in the variety, numbers and orders of their constituent amino acids linked by peptide bonds only.

2. The secondary structure is the shape that the polypeptide chain forms as a result of hydrogen bonding. This is most often a spiral known as the α helix. An alternative is a pleated sheet occurring as a flat zig-zag chain.

3. The tertiary structure is formed by the bending and twisting of the polypeptide helix into a compact structure. This gives the molecule its 3D shape. The shape is maintained by disulphide, ionic and hydrogen bonds.

4. The quaternary structure arises from a combination of two or more polypeptide chains in tertiary form. These are associated with non-protein groups and form large, complex molecules, e.g. haemoglobin.

Classification of proteins

Proteins can be divided into two groups according to their structure:

- Fibrous proteins perform structural functions. They consist of polypeptides in parallel chains or sheets with numerous cross-linkages to form long fibres. For example, keratin (in hair). Fibrous proteins are insoluble in water, strong and tough. Collagen provides tough properties needed in tendons. A single fibre consists of three polypeptide chains twisted around each other like a rope. These chains are linked by cross-bridges making a very stable molecule.

- Globular proteins perform a variety of different functions – enzymes, antibodies, plasma proteins and hormones. These proteins are compact and folded as spherical molecules. They are soluble in water. Haemoglobin consists of four folded polypeptide chains, at the centre of which is an iron-containing group called haem.

Inorganic ions

Inorganic ions play important roles in living organisms. In plants, mineral ions are transported dissolved in water. They can be divided into two groups:

- Macronutrients which are needed in small amounts. These include:
  - Magnesium, a constituent of chlorophyll in leaves.
  - Iron, a constituent of haemoglobin in blood.
  - Phosphate, found in the plasma membrane, nucleic acids, atp.
  - Calcium, a constituent of bones and teeth.
- Micronutrients which are needed in minute (trace) amounts, e.g. copper, zinc.

Water

Apart from providing a habitat for aquatic organisms, water plays an important role in plants and animals with key elements found in aqueous solution. Water is transparent, allowing light to pass through, enabling aquatic plants to photosynthesise effectively.

Water acts as a medium for metabolic reactions. Water makes up between 65% and 95% by mass of most plants and animals. It is an important constituent of cells. The hydrophobic property of lipids is important in cell membranes.

>> *Pointer*

You should be able to use given structural formulae (proteins, triglycerides and carbohydrates) to show how bonds are formed and broken by condensation and hydrolysis, including peptide, glycosidic and ester bonds.

 Classify the following proteins as fibrous or globular: insulin, collagen, keratin, lysozyme (an enzyme).

Grade boost

You are required to learn the functions of these four elements:
Magnesium
Iron
Phosphate
Calcium.

>> *Pointer*

Compare a polypeptide to a piece of string. In a fibrous protein several strands are twisted together like a rope, whereas in a globular protein the string is rolled into a ball.

Grade boost

A molecule gets its properties from its structure. Be prepared to explain how the properties of water enable it to carry out its many important roles in living organisms.

quickfire

⑪ Why is water described as a polar molecule?

quickfire

⑫ State the properties of water which allow the following:
How are insects able to walk on water?
How are fish able to live in a frozen pond?
Why does sweating keep us cool?

quickfire

⑬ Why are the following properties of water important to living organisms?
1. It is a universal solvent.
2. It is transparent.

Cohesion and surface tension

- Water is a polar molecule and has no overall charge. The oxygen end of the molecule has a slight negative charge and the hydrogen end of the molecule has a slight positive charge. When two water molecules are in close contact their opposing charges attract each other forming a hydrogen bond. Individually the hydrogen bonds are weak but because there are many of them they stick together in a strong lattice framework. This sticking together of water molecules is called cohesion. This means that tall columns of water can be drawn up xylem vessels in tall trees.

- At ordinary temperatures water has the highest surface tension of any liquid except mercury. In a pond the cohesion between water molecules produces surface tension so that the body of an insect such as the pond skater is supported.

Water as a solvent

- Because water is a polar molecule it will attract other charged particles, such as ions, and other polar molecules, such as glucose. This allows chemical reactions to take place in solution and since these chemicals dissolve in water, it acts a transport medium, e.g. in animals blood transports many dissolved substances. In plants water transports minerals in the xylem and sucrose in the phloem. Non-polar molecules such as lipids will not dissolve in water.

Thermal properties

- Water has a high specific heat. A large amount of heat energy is needed to raise the temperature of water. This is because the hydrogen bonds between water molecules restrict their movement. This prevents large fluctuations in the temperature of water and this is particularly important in keeping the temperature of aquatic habitats stable so that organisms do not have to endure extremes of temperature. This also allows enzymes within the cells to work effectively.

- Water has a high latent heat, i.e. a great deal of heat energy is needed to change it from a liquid to a vapour state. This is important, for example, in temperature control where heat is used for vaporisation of water when sweating. That is, the evaporation of water from a surface results in cooling.

Density

- Water has a maximum density at 4°C. Water in its solid form (ice) is less dense than water and so floats on the surface. Ice forms an insulating layer and allows organisms to survive beneath it.

Cell structure and organisation

Grade boost

Be prepared to draw and label a prokaryote cell and to make a comparison with an eukaryote cell.

Cell organisation

There are two types of cell: prokaryotic cells and eukaryotic cells. Prokaryote cells have a simple structure and were probably the first forms of life on Earth. Eukaryotic cells probably evolved from eukaryote cells around 1000 million years ago.

An example of a prokaryote cell is a bacterium. Eukaryote cells are typical of the great majority of organisms including all animals and plants.

Prokaryotic cells	Eukaryotic cells
Found in bacteria and blue-green algae	Found in plants, animals, fungi and protoctists
No membrane-bound organelles	Membrane-bound organelles
DNA lies free in the cytoplasm	DNA located on chromosomes
No nuclear membrane or ER	Distinct membrane-bound nucleus
Ribosomes are smaller	Ribosomes are larger
Cell wall containing murein	Cell wall in plants made of cellulose

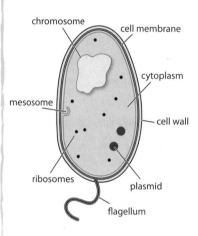

Structure of a generalised bacterial cell.

Viruses

Viruses cause a variety of infectious diseases in humans, animals and plants. Viruses are extremely small and can only be seen using an electron microscope. They can be called 'non-cells' as they have no cytoplasm, no organelles and no chromosomes. Outside a living cell a virus exists as an inert 'virion'. When they invade a cell they are able to take over the cell's metabolism and multiply within the host cell. Each virus particle is made up of a core of nucleic acid surrounded by a protein coat, the capsid. Most viruses are found in animal cells and those attacking bacteria (bacteriophages) have the nucleic acid DNA. Other animal and plant viruses contain RNA. A widely studied virus is T2 phage, a bacteriophage, which infects the bacterium *Escherichia coli (E.coli)*

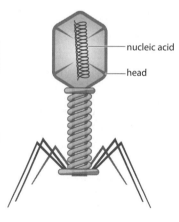

A typical virus.

⑭ What are the two major components of a virus?

Cell structure

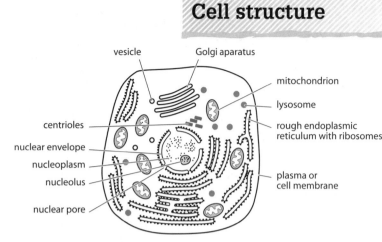

Generalised structure of an animal cell as seen using an electron microscope.

The cell is the basic unit of life and each cell can be thought of as a separate place where chemical processes of that cell take place. Simple organisms consist of only one cell, that is, they are unicellular. More advanced organisms consist of many cells and are said to be multicellular, where different cells are specialised to carry our particular functions. Plant and animal cells contain a number of organelles that perform a variety of functions. The electron microscope enables scientists to view the detailed structure of these organelles. This is known as the ultrastructure of the cell.

The cytoplasm is a highly organised material consisting of a soluble ground substance called the cytosol in which are found a variety of organelles.

Grade boost

You should be able to recognise organelles in electron micrographs.

Grade boost

Compare the cell structure of eukaryote, animal and plant, prokaryote and virus.

>> Pointer

The cell is a 3D structure. An electron micrograph may show mitochondria as circular or sausage-shaped as they are cut in different planes.

Nucleus

This is the most prominent feature in the cell. Its function is to control the cell's activities and to retain the chromosomes. The nucleus is bounded by a double membrane, the nuclear membrane (or envelope). This has pores in it to allow the transport of mRNA. The cytoplasm-like material within the nucleus is called the nucleoplasm. It contains chromatin, which is made up of coils of DNA (histones) bound to protein. During cell division the chromatin condenses to form the chromosomes. Within the nucleus is a small spherical body called a nucleolus. Its function is to manufacture RNA which is needed to make ribosomes.

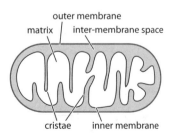

Basic structure of a mitochondrion.

Mitochondrion

The mitochondrion has a double membrane separated by a narrow fluid-filled inter-membrane space. The inner membrane is folded inwards to form extensions called cristae. The interior of the mitochondrion contains an organic matrix containing numerous chemical compounds. Mitochondria are the sites of aerobic respiration in the cell. Some of the reactions take place in the matrix while others occur on the inner membrane. The cristae increase the surface area on which the respiratory processes take place. The function of mitochondria is to produce energy as ATP. Muscle cells contain large numbers of mitochondria, reflecting the high metabolic activity taking place there.

Endoplasmic reticulum (ER)

This consists of an elaborate system of parallel double membranes forming flattened sacs. The fluid-filled spaces between the membranes are called cisternae. The ER is connected with the nuclear membrane and may link to the Golgi body. The cavities are interconnected and this system allows the transport of materials throughout the cell. There are two types of ER:

- Rough ER has ribosomes on the outer surface. The rough ER functions in transporting proteins made by the ribosomes. Rough ER is present in large amounts in cells that make enzymes that may be secreted out of the cell.
- Smooth ER has membranes which lack ribosomes. These are concerned with the synthesis and transport of lipids.

Ribosomes

These are made up of one large and one small sub-unit. They are manufactured in the nucleolus from ribosomal RNA and protein. They are important in protein synthesis.

Golgi body

This is similar in structure to ER but is more compact. The Golgi body is formed by rough ER being pinched off at the ends to form small vesicles. A number of these vesicles then fuse together to form the Golgi body. Proteins are transported in the vesicles and are modified and packaged in the Golgi body. For example, proteins may be combined with carbohydrates to make glycoproteins. At the other end of the Golgi body vesicles can be pinched off and the products secreted by **exocytosis** when the vesicle moves to and fuses with the cell membrane.

Other functions of the Golgi body include:

- Producing secretory enzymes.
- Secreting carbohydrates, e.g. for the formation of plant cell walls.
- Producing glycoprotein.
- Transporting and storing lipids.
- Forming lysosomes.

Key terms

Exocytosis = when a substance leaves a cell after being transported through the cytoplasm in a vesicle.

Phagocytosis = the process by which the cell obtains solid materials that are too large to be taken in by diffusion.

Grade boost

Membranous organelles are enclosed areas within the cytoplasm. This has advantages in that potentially harmful chemicals and/or enzymes can be isolated. Membranes also provide a large surface area for the attachment of enzymes involved in metabolic processes, as well as providing a transport system within the cell.

Pointer

Once you have used the term 'endoplasmic reticulum' it is acceptable to abbreviate to 'ER'.

quickfire

⑮ The following are functions of different organelles. Name the organelle in each case:
1. Protein synthesis.
2. Producing glycoproteins.
3. Producing ATP.
4. Producing ribosomes.

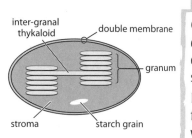

Basic structure of a chloroplast.

inter-granal thykaloid
double membrane
granum
stroma
starch grain

Lysosomes

Lysosomes are small vacuoles formed when portions of the Golgi body are pinched off. They contain and isolate digestive enzymes from the remainder of the cell. They can also release these enzymes and destroy worn out organelles in the cell. Digestion is carried out in the membrane-lined vacuole into which several lysosomes may discharge their contents.

They can also digest material that has been taken into the cell, e.g. white blood cells engulf bacteria by **phagocytosis** and the lysosomes discharge their contents into the vesicle so formed and digest the bacterium.

Centrioles

Centrioles are found in all animal cells but are absent from the cells of higher plants. Centrioles are located just outside the nucleus in a distinct region of the cytoplasm known as the centrosome. A centriole consists of two hollow cylinders positioned at right angles to one another. During cell division centrioles divide and move to opposite poles of the cell where they synthesise the microtubules of the spindle.

Chloroplast

Chloroplasts are found in the cells of photosynthesising tissue. They have a double membrane inside which is the fluid stroma, containing ribosomes, lipid, circular DNA and possibly starch. Within the stroma are number of flattened sacs called thylakoids. A stack of thylakoids is called a granum.

Each granum consists of between two and a hundred of these closed, parallel, flattened sacs. The photosynthetic pigments such as chlorophyll are found within the thylakoids. This arrangement produces a large surface area for trapping light energy.

Vacuole

Plant cells have a large permanent vacuole which consists of a fluid-filled sac bounded by a single membrane, the tonoplast. Vacuoles contain cell sap, a storage site for chemicals such as glucose, and provide an osmotic system which functions in support of young tissues.

Animal cells contain vacuoles but these are small, temporary vesicles and may occur in large numbers.

Cellulose cell wall

The cell wall consists of cellulose microfibrils embedded in a polysaccharide matrix. The main functions of a cell wall are:

- To provide strength and support.
- To permit the movement of water from cell to cell.

Differences between plant and animal cells

Plant cells have all the structures found in animal cells plus some additional features:

Plant cells	Animal cells
Cell wall	No cell wall
Chloroplasts	No chloroplasts
Large permanent vacuole	Small, temporary vacuoles
No centriole	Centriole
Plasmodesmata	**No plasmodesmata**

Levels of organisation

Unicellular organisms carry out all life functions within a single cell.

Multicellular organisms are more efficient as they contain a variety of different cells. As they develop, each cell becomes specialised in structure to suit the role it will perform. This is called cell differentiation. For example, nerve cells become long and thin to carry impulses. Some cells remain undifferentiated and function as 'packing' cells, for example parenchyma cells in plants.

Tissues

Cells are usually grouped together. A tissue consists of a collection of similar cells that carry out a particular function, for example epithelial tissue. These are sheets of cells that line the surface of organs in animals. These cells often have a protective or secretory function.

Just as cells are grouped into tissues, different tissues are aggregated into organs.

- An organ is composed of several different tissues that are coordinated to perform a function, for example the eye.
- Organs work together as a single unit or organ system, for example the digestive system.
- Consequently organisms are made up of a number of different systems that work together.

Key Terms

You will come across these terms later in the course.

Centriole = structure in cell division from which spindle fibres develop.

Plasmodesmata = strands of cytoplasm linking cells together.

quickfire

⑰ List three features present in a plant cell, not found in an animal cell.

quickfire

⑱ State whether each of the following is a cell, tissue or organ: kidney, epithelium, muscle, sperm.

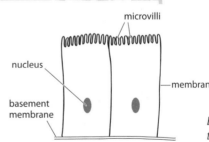

Epithelial cells from the lining of the small intestine.

Grade boost

Be prepared to explain why Singer and Nicholson called their theory the Fluid Mosaic model.

⑲ What is the difference between extrinsic and intrinsic proteins found in the cell membrane?

Grade boost

Although all plasma membranes have the same phospholipid bilayer structure, they have different properties according to the different substances they contain, particularly proteins.

Cell surface membrane

Membrane structure

The cell surface membrane or plasma membrane is the boundary that separates the living cell from its non-living surroundings. The membrane also controls which substances pass into and out of the cell. The cell membrane is made up almost entirely of proteins and phospholipids.

The phosphate head of the phospholipid is a polar molecule (hydrophilic or water-loving) and has an attraction for other polar molecules such as water. The fatty acid end of the phospholipid, made up of two fatty acid tails, is non-polar (hydrophobic or water-hating) and repels water. Phospholipids can form bilayers with one sheet of phospholipid forming over another. This phospholipid bilayer forms the basis of membrane structure. The use of the electron microscope enabled Singer and Nicholson to put forward their theory called the Fluid Mosaic model in 1972. They proposed that:

- There is a **bimolecular** phospholipid layer.
- Associated with the bilayer is a variety of protein molecules:
  - Extrinsic proteins which occur on the surface of the bilayer or partly embedded in it.
  - Intrinsic proteins which extend across both layers.
- The phospholipid layer is capable of movement, i.e. it is fluid and in surface view the proteins are dotted throughout the layer in a mosaic arrangement.

The model is referred to as the 'Fluid Mosaic' model because the components are free to move with respect to each other and the proteins embedded in the bilayer vary in shape, size and pattern.

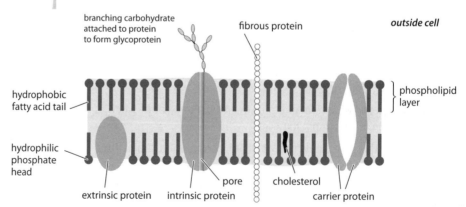

Fluid Mosaic model of the cell membrane.

Cholesterol is also found in animal cells. It fits between the phospholipid molecules, increasing the rigidity and stability of the membrane. Glycolipids (lipids that have combined with polysaccharide) are also found in the outer layer of the membrane and are thought to be involved in cell-to-cell recognition. Glycoproteins also stick out of some membranes.

The main functions of the proteins in the cell membrane include:

- Oroviding structural support.
- Allowing active transport across the membrane by forming ion channels.
- Forming recognition sites by identifying cells.
- Acting as carriers transporting water-soluble substances across the membrane.

The membrane as a barrier

Lipid-soluble substances move through the membrane via the phospholipid part. It prevents the entry or exit of water-soluble substances. The latter pass through special protein molecules, which form water-filled channels across the membrane. The cell surface membrane is selectively permeable to water and some solutes. Lipid-soluble substances can move through the cell membrane more easily than water-soluble substances.

- Small, uncharged molecules, such as oxygen and carbon dioxide, freely pass through the membrane as they are soluble in the lipid part.
- Lipid soluble molecules such as glycerol can pass through the membrane.
- The hydrophobic core of the membrane impedes the transport of ions and polar molecules.
- Charged particles (ions) and relatively large molecules such as glucose cannot diffuse across the non-polar centre of the phospholipid bilayer because they are relatively insoluble in lipid. Intrinsic proteins assist such particles to pass in or out of the cell by a passive process called facilitated diffusion.

Key Term
Passive transport = does not involve the use of energy.

quickfire

20 Glucose is water soluble and vitamin A is fat soluble. They both enter a cell by passing across the membrane. Explain how the properties of the molecules and the structure of the membrane determine the way in which these two substances pass across.

Grade boost

This is a difficult concept. The hydrophobic core of the membrane impedes the transport of ions and polar molecules. These require specific proteins to help them across.

Transport across cell membranes

Diffusion

Diffusion is the movement of molecules or ions from a region where they are in high concentration to a region of lower concentration until they are equally distributed. Ions and molecules are always in a state of random movement, but if they are highly concentrated in one area, there will be a net movement away from that area until equilibrium is reached or until there is a uniform distribution.

Pointer

Using the term 'short diffusion path' in the correct context will impress.

Grade boost

'Concentration gradient' is a useful term. Molecules move down a concentration gradient from high to low.

>> **Pointer**

Diffusion is proportional to the difference in concentration between two areas. It is incorrect to state that it is proportional to concentration.

quickpire

㉑ State two features of the membrane that increase the rate of diffusion.

quickpire

㉒ Explain how an increase in temperature affects the rate of diffusion.

>> **Pointer**

Facilitated diffusion is a special form of diffusion that allows faster movement of molecules.

It is a passive process and occurs down a concentration gradient. However, it occurs at specific points on the plasma membrane where there are special protein molecules.

quickpire

㉓ How does facilitated diffusion differ from diffusion?

The rate of diffusion is affected by:

- The concentration gradient, i.e. the greater the difference in the concentration of molecules in two areas, the greater the rate.
- The distance of travel, i.e. the shorter the distance between two areas the greater the rate.
- The surface area of the membrane – the larger the area the quicker the rate.
- The thickness of the membrane – the thinner the membrane the greater the rate.
- An increase in temperature results in an increase in rate, since there is an increase in molecular energy and therefore movement.

Facilitated diffusion

Charged particles or ions and large molecules such as glucose do not readily pass through the cell membrane because they are relatively insoluble in lipid. In the cell membrane, protein molecules span the membrane from one side to the other and help such particles to diffuse in to or out of the cells. These proteins are of two types:

- **Channel proteins** – consist of pores lined with polar groups allowing charged ions to pass through. (As the channel is hydrophilic, water-soluble substances can pass through.) As each channel protein is specific for one type of ion each protein will only let one particular ion through. They can also open and close according to the needs of the cell.
- **Carrier proteins** – allow the diffusion across the membrane of larger polar molecules such as sugars and amino acids. A particular molecule attaches to the carrier protein at its binding site and causes the carrier protein to change its shape, releasing the molecule through the membrane.

Carrier proteins and channel proteins increase the rate of diffusion along the concentration gradient without the need for energy in the form of ATP from respiration.

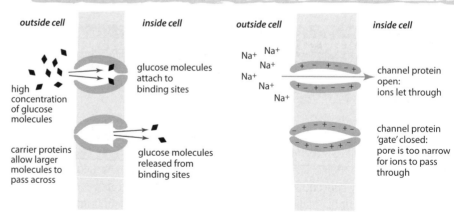

Carrier and channel proteins.

Active transport

Unlike the processes described so far, active transport is an energy-requiring process in which ions and molecules are moved across membranes against a concentration gradient.

The features of active transport are:

- Ions and molecules can move in the opposite direction to that in which diffusion occurs, i.e. against a concentration gradient.
- The energy for active transport is supplied by ATP, and anything which affects the respiratory process will affect active transport.
- Active transport will not take place in the presence of a respiratory inhibitor such as cyanide.
- The process occurs through the carrier proteins that span the membrane. The proteins accept the molecule and then the molecule enters the cell by a change in shape of the carrier molecule.

Processes involving active transport include protein synthesis, muscle contraction, nerve impulse transmission, absorption of mineral salts by plant roots.

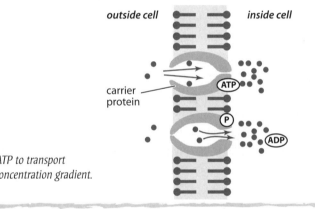

Carrier proteins use ATP to transport molecules against a concentration gradient.

>> **Pointer**

The exchange of substances between cells and their surroundings occur in ways that involve metabolic energy (active transport) and in ways that do not (passive transport).

Passive processes require no external energy, only the kinetic energy of the molecules themselves.

㉔ State one similarity and one difference between facilitated diffusion and active transport.

㉕ Name two processes that involve active transport.

Graph showing effect of respiratory inhibitor on the rate of uptake of a substance across a cell membrane

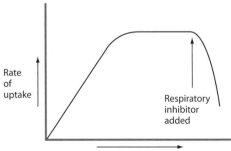

>> **Pointer**

The graph shows that at higher concentration differences, a plateau is reached when the carrier proteins are saturated. The rate of uptake is affected with the addition of a respiratory inhibitor. Active transport must be taking place as the process needs ATP.

Solute = any substance that is dissolved in a solvent. Solutes and solvents form a solution.

Osmosis = the passage of water from a region of high water potential to a region of lower water potential through a partially permeable membrane.

≫ Pointer

Membranes are partially permeable, that is, they are permeable to water molecules and some other small molecules but not to larger molecules. Osmosis is a specialised form of diffusion involving only water molecules.

Grade boost

Water potential is the pressure created by water molecules and is measured in kilopascals (kPa). Pure water has a water potential of zero. The addition of a solute to pure water will lower the water potential. That is the solution has a negative value.

Grade boost

Be prepared to use the given equation in calculations.

quickfire

㉖ The water potential of three cells is:
Cell A = −150 kPa
Cell B = −200 kPa
Cell C = −100 kPa
Place the cells in the order that water would pass from one cell to the next.

Osmosis

Most cell membranes are permeable to water and certain **solutes** only. In biological systems **osmosis** is a special form of diffusion which involves the movement of water molecules only.

Biologists use the term water potential (WP) ψ(psi) to describe the tendency of water molecules to leave a system.

Pure water has the highest water potential of zero. This is because where there is a high concentration of water molecules, they have a greater potential energy, i.e. the water molecules are completely free to move about. When a solute, such as sugar, is dissolved in water, there are proportionally fewer water molecules to move about and the water potential of the solution is lowered. All water potentials (except that of pure water) have a negative value. The more concentrated the solution, the more negative the water potential, i.e. the fewer free water molecules there are.

In plant cells the following equation is used to describe the relationship between the forces:

$$\psi \quad = \quad \psi_s \quad + \quad \psi_p$$

water potential = solute potential + pressure potential

- The presence of solute molecules in the vacuole of a plant cell lowers the WP.
- The concentration of dissolved substances inside the cell vacuole is called the solute potential.
- When water enters a plant cell vacuole by osmosis, a hydrostatic pressure is set up and pushes outwards on the cell wall. As the outward pressure builds up, the cell wall develops an opposing force called the pressure potential. The pressure potential is usually positive.

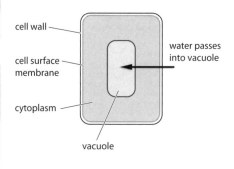
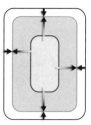

cell wall

cell surface membrane

cytoplasm

vacuole

water passes into vacuole

as water enters the cell, the rigid cell wall develops an opposing pressure potential

Solute and pressure potential.

Turgor and plasmolysis

- If the WP of the external solution is lower than the solution inside the cell, it is said to be hypertonic and water flows out of the cell.
- If the WP of the external solution is higher than the solution inside the cell, it is said to be hypotonic and water flows into the cell.
- If the cell has the same solute concentration as the surrounding solution, the external solution is isotonic with that of the cell.
- When a plant cell is placed in a hypertonic solution, it loses water by osmosis. The vacuole shrinks and the cytoplasm will draw away from the cell wall. This process is called plasmolysis and when complete, the cell is said to be flaccid.
- The point at which the cell membrane *just* begins to move away from the cell wall is said to be the point of incipient plasmolysis.
- A plant cell will gain water if placed in a hypotonic solution and will continue to take in water until prevented by the opposing wall pressure. The pressure potential rises until it is equal and opposite to the solute potential. In theory the water potential is now zero and when the cell cannot take in any more water it is said to be **turgid**. The state of turgor is important in plants, particularly young seedlings. It supports them and maintains their shape and form.

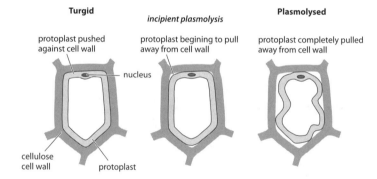

Turgid

protoplast pushed against cell wall

— nucleus

cellulose cell wall — protoplast

incipient plasmolysis

protoplast begining to pull away from cell wall

Plasmolysed

protoplast completely pulled away from cell wall

Turgid and plasmolysed cells.

As it has no cell wall, an animal cell is affected differently. If red blood cells are placed in distilled water, water enters by osmosis and it bursts. This is called haemolysis. If red blood cells are placed into strong salt solution, water passes out of the cell and the cell shrinks.

quickfire

㉗ What is meant by the term 'plasmolysis'?

》 Pointer

Turgor plays an important role in young seedlings and herbaceous (non-woody) plants. It helps support them and maintains their shape and form.

》 Pointer

Don't forget to include any practical work connected with water potential and plasmolysis of cells in your revision.

Grade boost

Explain why it is essential for the saline (salt) drip used in hospitals to be isotonic with the body cells.

>> *Pointer*

In cells, metabolic reactions take place quickly and thousands of reactions are taking place simultaneously. Order and control are essential if reactions are not to interfere with each other. These features of metabolism are made possible by the action of enzymes.

>> *Pointer*

Induced fit model of enzyme action – this theory is a modified version of the lock and key hypothesis and proposes that the enzyme changes its shape slightly to accommodate the substrate. As the enzyme changes its shape it places a strain on the substrate molecule and distorts a particular bond. This lowers the activation energy needed to break the bond.

Grade boost

It is incorrect to state that the substrate has the 'same shape' as the active site. It has a complementary shape to the active site.

Enzymes

Enzyme structure

Enzymes are tertiary globular proteins where the protein chain is folded back on itself into a spherical or globular shape. Each enzyme has its own sequence of amino acids and is held in its tertiary form by hydrogen bonds, disulphide bridges and ionic bonds. This complex 3D shape gives the enzyme many of its properties. Although the enzyme molecule is large, only a small region, called the active site, is functional.

How enzymes work

Enzymes are biological catalysts that speed up the rate of metabolic reactions. These reactions can be of two types:

1. Reactions where larger molecules are broken down into smaller molecules.
2. Reactions where small molecules are built up into larger, more complex, molecules.

Enzymes react with another molecule called a substrate. Each enzyme has its own special shape, with an area, the active site, onto which the substrate molecules bind.

Enzyme + substrate = enzyme-substrate complex = enzyme + product.

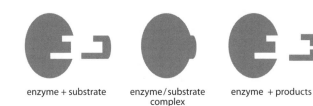

enzyme + substrate   enzyme/substrate complex   enzyme + products

Enzyme and substrate complex.

Properties of enzymes

- Enzymes are specific, i.e. each enzyme will catalyse only one particular reaction, for example sucrase acts on the sugar, sucrose.
- Enzymes are very efficient and have a high turnover number. This means that they can convert many molecules of substrate per unit time; for example, catalase, which breaks down the waste product hydrogen peroxide in the body, has a turnover number of several million!
- Chemical reactions need energy to start them off and this is called activation energy. This energy is needed to break the existing chemical bonds inside molecules. In the body, enzymes lower the activation energy of a reaction and so reduce the input of energy needed and allow reactions to take place at lower temperatures.

Factors affecting the rate of enzyme action

Enzymes are made inside living cells but may act inside the cell (intracellular) or outside (intercellular, extracellular) such as the digestive enzymes of the alimentary canal. Environmental conditions, such as temperature and pH, change the three dimensional structure of enzyme molecules. Bonds are broken and hence the configuration of the active site is altered.

⌃⌃ Grade boost

Enzymes are inactive at 0°C and if the temperature is raised they become active again. Enzymes are denatured at temperatures above 40°C.

Temperature

An increase in temperature gives molecules greater kinetic energy and they move around more quickly, increasing the chance of molecules colliding. Increasing the temperature of an enzyme-controlled reaction results in an increase in the rate of reaction. As a general rule, the rate of reaction doubles for each 10°C rise in temperature until an optimum temperature is reached. For most enzymes this is 40°C.

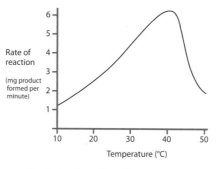

Graph showing effect of temperature on rate of reaction.

Above this temperature, the increasing vibration of the molecules causes the hydrogen bonds to break, causing a change in the tertiary structure of the enzyme. This alters the shape of the active site and the substrate will not fit into the active site. The enzyme is then said to be denatured. This is a permanent change in the structure. If enzymes are subjected to low temperatures, such as freezing, the enzyme is inactivated as the molecules have no kinetic energy. However, the enzyme can work again if the temperature is raised.

pH

The rate of an enzyme-catalysed reaction will vary with changes in pH. Enzymes have a narrow optimum range and small changes in pH can affect the rate of reaction without affecting the structure of the enzyme. Small changes in pH outside the optimum can cause small reversible changes in enzyme structure and result in inactivation. Extremes of pH can denature an enzyme.

The charges on the amino acid side-chains of the enzyme's active site are affected by free hydrogen ions or hydroxyl ions. In the formation of an enzyme substrate complex the charge on the active site must match those on the substrate. If the active site has too many H^+ ions (say) the active site and the substrate may both have the same charge and the enzyme will repel the substrate.

◉ ⫷⫷⫷ quickfire

㉘ Describe the effect of an increase in temperature from 0°C to 40°C on the rate of an enzyme-controlled reaction.

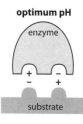

optimum pH

enzyme

substrate

charges on active site match those of substrate so an enzyme-substrate complex forms

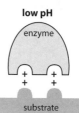

low pH

enzyme

substrate

charges on active site repel substrate

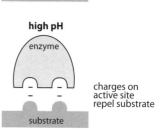

high pH

enzyme

substrate

charges on active site repel substrate

Effect of pH.

At extremes of pH the hydrogen bonding is affected and the three-dimensional shape of the enzyme is altered and so is the shape of the active site.

Enzymes are also affected by the concentration of the substrate and the concentration of the enzyme itself.

Substrate concentration

The rate of an enzyme-catalysed reaction will vary with changes in substrate concentration. If the amount of enzyme is constant, the rate of reaction will increase as the substrate increases. At low substrate concentrations the enzyme molecules have only a limited number of substrate molecules to collide with. In other words the active sites are not working to full capacity. As more substrate is added, there must come a point when all the enzyme's active sites are filled. In other words, the rate of reaction is at a maximum.

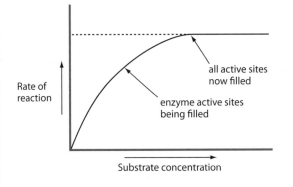

Graph showing effect of increasing substrate on rate of reaction.

Enzyme concentration

The rate of an enzyme-catalysed reaction will vary with changes in enzyme concentration. Increasing the enzyme concentration will increase the rate of reaction.

In enzyme experiments it is essential that buffers and controls are used:

- *Buffers* – buffers maintain a constant pH. When a buffer is used in an experiment, the pH changes little when a small quantity of acid or alkali is added. It may be said that a buffer 'soaks up hydrogen ions'.

- *Controls* – controls are duplicate experiments, identical in every respect to the actual experiment, except the variable being investigated, which is kept constant. For example, boiled enzyme may be used in a control experiment instead of the enzyme.

Enzyme inhibition

Inhibition occurs when enzyme action is slowed down or stopped by another substance. The inhibitor combines with the enzyme and either directly or indirectly prevents it forming an enzyme-substrate complex.

Competitive inhibition

The inhibitor is structurally similar to the substrate and competes with the active site of the enzyme, i.e. the inhibitor has a shape that lets it fit into the active site of the enzyme in place of the substrate. For example, malonic acid competes with succinate for the active sites of succinic dehydrogenase, an important enzyme in the Krebs cycle in respiration. If the substrate concentration is increased, it will reduce the effect of the inhibitor. This is because the more substrate molecules present, the greater the chance of finding active sites, leaving fewer to be occupied by the inhibitor.

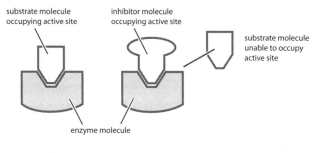

Competitive inhibition.

Non-competitive inhibition

The inhibitor binds to the enzyme at a site away from the active site. This alters the overall shape of the enzyme molecule, including the active site, in such a way that the active site can no longer accommodate the substrate. As the substrate and inhibitor molecules attach to different parts of the enzyme, they are not competing for the same sites. The rate of reaction is therefore unaffected by substrate concentration. For example, cyanide (a respiratory poison) attaches itself to part of the enzyme cytochrome oxidase, and inhibits respiration.

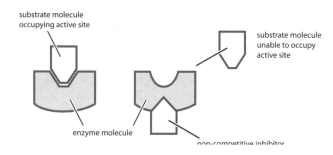

Non-competitive inhibition.

Grade boost

Describe enzyme–substrate reactions in terms of molecular collisions. With competitive inhibition the greater the substrate concentration compared to the inhibitor, the greater the chance that the substrate will collide with the enzyme.

quickfire

(31) State how the two types of inhibitors differ in how they attach to an enzyme.

quickfire

(32) In an experiment an enzyme-controlled reaction is inhibited by substance A. Suggest the experimental change which could be made to determine whether the inhibitor is competitive or non-competitive.

Grade boost

Do not be too concerned with the functions of malonic acid and cytochrome oxidase at this stage. These chemicals and their involvement in respiration may be studied at A2.

Medical and industrial applications of enzymes

Immobilised enzymes

These are enzyme molecules that are fixed, bound or trapped on an inert matrix such as a gel capsule (alginate beads). These beads can be packed into glass columns. Substrate can be added to the top of the column and it reacts with the enzyme as it slowly flows down the column. Once set up, the column can be used again

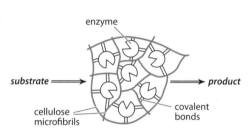

Immobilised enzymes in a framework of cellulose microfibrils.

and again. As the enzyme is fixed it does not get mixed up with the products and is therefore cheaper to separate. Immobilised enzymes are used widely in industrial processes, such as fermentation, as they can readily be recovered for reuse.

Enzyme instability is one of the key factors that prevent the wider use of 'free' enzymes. Chemicals such as organic solvents, raised temperatures and pH values outside the norm can denature the enzyme with a consequent loss of activity. Immobilising enzymes with a polymer matrix creates a microenvironment allowing reactions to occur at higher temperatures than normal. This means that activity is increased and so production is also increased.

Other advantages include:

- Enzymes can tolerate a wider range of conditions.

- Enzymes are easily recovered for re-use thus reducing overall costs.

- Several enzymes with differing pH or temperature optima can be used together.

- Enzymes can be easily added or removed giving greater control over the reaction.

Grade boost

You should be able to list the advantages of immobilised enzymes.

③③ Using the graph, list three differences between the effects of temperature on the immobilised and free enzyme.

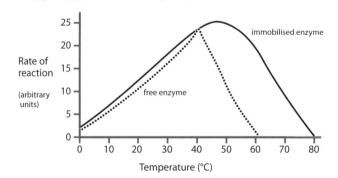

Graph showing effect of temperature on free and immobilised enzymes.

Biosensors

One use of immobilised enzymes involves **biosensors**, which work on the principle that enzymes are specific and are able to select one type of molecule from a mixture even in very low concentrations. A biosensor can be used in the rapid and accurate detection of minute traces of biologically important molecules.

Biosensors have great potential in the areas of medical diagnosis and environmental monitoring. The electrode probe can detect changes in substrate or product, temperature changes or optical properties.

One particular use of a biosensor is in the detection of blood sugar in diabetics. The electrode probe, which has a specific enzyme immobilised in a membrane, is placed in the blood sample. If glucose is present, it diffuses through the membrane, and forms an enzyme-substrate complex. The reaction produces a small electric current, which is picked up by the electrode (the transducer). This current is read by a meter which produces a reading for blood glucose. Normal blood glucose levels are described as being 3.89–5.83 mmol dm^{-3}.

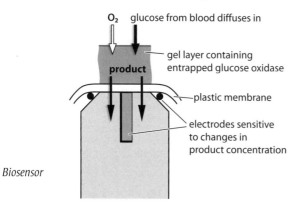

Biosensor

Steps in using a biosensor:

1. Blood contains a mixture of different molecules.
2. Enzyme electrode is placed in a blood sample.
3. Glucose diffuses into the immobilised enzyme layer.
4. Oxygen is taken up.
5. The rate of oxygen uptake is proportional to the glucose concentration.
6. A digital display shows an accurate concentration of glucose.

Key Term

Biosensor = describes the association of a biomolecule, such as an enzyme, with a transducer, which produces an electrical signal in response to substrate transformation. The strength of the electrical signal may be measured with a meter.

≫ Pointer

Immobilised enzymes are also used in pregnancy testing kits and in a fermenter to provide a rapid, sensitive and specific measurement of products.

quickfire

㉞ The Benedict's test is at best semi-quantitative estimating the approximate concentration of glucose in a sample. State two advantages of using a biosensor.

Polymer = a large number of repeating units.

>> *Pointer*

DNA performs two major functions:

- Replication in dividing cells.
- Carrying the information for protein synthesis.

These processes are included in the A2 specification.

>> *Pointer*

DNA determines inherited characteristics and contains information in the form of the genetic code.

>> *Pointer*

Bases that pair together are called complementary base pairs. A pairs with T, and C pairs with G. In an exam answer do not use abbreviations!

>> *Pointer*

Nitrogenous bases derived from pyrimidines are single ring structures, whereas those derived from purines are double ring structures.

Grade boost

Cytosine and thymine include the letter 'Y', so does pyrimidine!

Nucleic acids

There are two types of nucleic acid: deoxyribonucleic acid (DNA) and ribonucleic acid (RNA). Both are built up of units called nucleotides. Individual nucleotides are made up of three parts that combine by condensation reactions. These are:

- Phosphate ion. This has the same structure in all nucleotides.
- Pentose sugar, of which there are two types:
    1. In ribonucleic acid (RNA) the sugar is ribose.
    2. In deoxyribonucleic acid (DNA) the sugar is deoxyribose.
- Organic base which contains nitrogen.

There are five different bases which are divided into two groups:

- The pyrimidine bases (single ring structures) are thymine, cytosine and uracil.
- The purine bases (double ring structures) are adenine and guanine.

The structure of DNA

- DNA is a double-stranded **polymer** of nucleotides or polynucleotide.
- Each polynucleotide may contain many million nucleotide units.
- It is in the form of a double helix, the shape of which is maintained by hydrogen bonding.
- The pentose sugar is always deoxyribose.
- DNA contains four organic bases. These are adenine, guanine, cytosine, and thymine.
- Each strand is linked to the other by pairs of organic bases.
- Cytosine always pairs with guanine, adenine always pairs with thymine, and the bases are joined by hydrogen bonds.
- Put simply, DNA is like a coiled ladder with the uprights of the ladder being made up of alternating sugars and phosphate groups and the rungs made up of the bases. The bases are held together by weak hydrogen bonds.

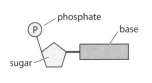

Nucleotide

Part of a DNA chain showing a polynucleotide consisting of three base pairs.

How DNA is well suited to carry out its functions:

- It is a very stable molecule and can pass from generation to generation without change.
- It is a very large molecule and can carry a large amount of genetic information.
- The two strands are able to separate easily as they are held together by weak hydrogen bonds.
- As the base pairs are held within the deoxyribose-phosphate backbone, the genetic information is protected from outside forces.

Grade boost

Questions may be asked about the stability of DNA.

》 Pointer

At AS you are not required to have a knowledge of protein synthesis.

The structure of RNA

- RNA is a single-stranded polymer of nucleotide.
- RNA contains the pentose sugar, ribose.
- RNA contains the organic bases adenine, guanine, cytosine, and uracil (in place of thymine).

There are three types of RNA and all are involved in the process of protein synthesis:

- Messenger RNA (mRNA) is a long single-stranded molecule formed into a helix. It is manufactured in the nucleus and carries the genetic code from the DNA to the ribosomes in the cytoplasm.
- Ribosomal RNA (rRNA) is found in the cytoplasm and is a large, complex molecule made up of both double and single helices. Ribosomes are made up of ribosomal RNA and protein. It is the site of translation of the genetic code.
- Transfer RNA (tRNA) is a small single-stranded molecule. It forms a clover-leaf shape, with one end of the chain ending in a cytosine-cytosine-adenine sequence at which point the amino acid it carries attaches itself. At the opposite end of the chain is a sequence of three bases called the anticodon. tRNA molecules transport amino acids to the ribosome so that proteins can be synthesised.

Comparing DNA and RNA

	DNA	**RNA**
Sugar	Deoxyribose	Ribose
Bases	CGAT	CGAU
Helix	Double	Single

㉟ State which type of RNA is found in the cytoplasm only and which type can be found in the nucleus and cytoplasm.

㊱ State two differences between DNA and RNA.

Cell division

Genetic information is copied and passed on to daughter cells.

Chromosome structure

Chromosomes are made up of DNA, protein and a small amount of RNA. DNA occurs as a single strand in the form of a double helix running the length of the chromosome. Each DNA molecule is made up of many sections called genes. It is only at the onset of cell division that chromosomes become visible. Shortly before cell division begins, each DNA molecule makes a copy of itself. The single thread of DNA becomes two identical threads. These are called chromatids and they lie parallel along most of their length but are joined only in a specialised region called the centromere.

Mitosis

Mitosis produces two daughter cells that are genetically identical to the parent cell.

Dividing cells undergo a regular pattern of events known as the cell cycle. This is a continuous process but for convenience of description it is subdivided into four stages preceded by a period when the cell is not dividing. This is called interphase and is the longest part of the cycle.

During interphase a newly formed cell increases in size and produces organelles lost during the previous division. The amount of DNA is also doubled. Just before the next cell division the chromosomes replicate so that each then consists of two chromatids joined together by the centromere. There is considerable metabolic activity as these processes need energy in the form of ATP. The chromosomes are not visible at interphase because the chromosome material, chromatin, is dispersed throughout the nucleus.

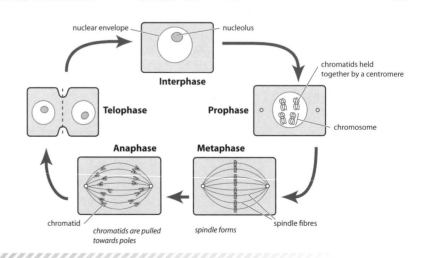

Stages in mitosis.

Stages in mitosis

Prophase – The chromosomes condense (shorten and become thicker) and become visible as long thin threads. They are now referred to as pairs of chromatids. In cells where centrioles are present, i.e. animals and lower plants, the centrioles divide and move to the opposite ends (poles) of the cells. Protein microtubules form from each centriole and the spindle develops, extending from pole to pole. Towards the end of prophase the nuclear membrane disintegrates and the nucleolus disappears. Pairs of chromatids can clearly be seen lying free in the cytoplasm.

Metaphase –The chromosomes arrange themselves at the centre or equator of the spindle and become attached to certain spindle fibres at the centromere. Contraction of these fibres draws the individual chromatids slightly apart.

Anaphase – A very rapid stage. The centromere splits and the spindle fibres contract and pull the now separated chromatids to the poles, centromere first.

Telophase –The chromosomes have now reached the poles of the cells and are referred to as chromosomes again. They uncoil and lengthen. The spindle breaks down, the centrioles replicate, the nucleoli reappear and the nuclear membrane reforms. In animal cells cytokinesis occurs by the constriction of the centre of the parent cell from the outside inwards. In plant cells, a cell plate forms across the equator of the parent cell from the centre outwards and a new cell wall is laid down.

Significance of mitosis

- Mitosis produces two cells that have the same number of chromosomes as the parent cell and each chromosome is an exact replica of one of the originals. The division allows the production of cells that are genetically identical to the parent and so gives genetic stability.

- By producing new cells, mitosis leads to growth of an organism and also allows for repair of tissues and the replacement of dead cells. An example of mitosis in plants is in the root tip. In human skin, dead surface cells are replaced by identical cells from below.

- Asexual reproduction results in complete offspring that are identical to the parent. This takes place in unicellular organisms such as yeast and bacteria. It also takes place in certain flowering plants where organs such as bulbs, tubers and runners produce large numbers of identical offspring in a relatively short period of time. There is no variation between each individual. However, most of these plants also reproduce sexually.

quickfire

㊲ Interphase is a period of intense metabolic activity. State three events that occur during this stage.

Grade boost

Mitosis is a continuous process. Under the microscope you will see only a snapshot of what is happening.

Grade boost

Be prepared to recognise the stages of mitosis from drawings and photographs and explain the events occurring during each stage. You may also be required to draw and label a particular stage.

quickfire

㊳ With particular reference to plants, what is the significance of mitosis?

》Pointer

You are not required to describe the complete process of meiosis but to describe the significance of the differences between mitosis and meiosis.

Meiosis

In sexual reproduction two gametes fuse to form a zygote. For each generation to maintain a full set of chromosomes (diploid number) the chromosome number must be halved (haploid number) during meiosis. Meiosis involves two divisions: meiosis I where the chromosome number is halved, and meiosis II where the two haploid nuclei divide again in a division identical to that of mitosis. The end result is the production of four daughter nuclei, each with half the number of chromosomes of the parent cell.

However, there is another important difference between meiosis and mitosis. During prophase I homologous chromatids wrap around each other and then partially repel each other but remain joined at certain points called chiasmata. At these points chromatids may break and recombine with a different but equivalent chromatid. This swapping of pieces of chromosomes is called crossing over and is a source of genetic variation.

During the first stage of metaphase of meiosis the pairs of **homologous** chromosomes arrange themselves randomly on the equator of the spindle. Only one of each pair passes into the daughter cell and this happens with each pair. Therefore the combination of chromosomes that goes into the daughter cell at meiosis I is also random. This random distribution and consequent independent assortment of chromosomes produces new genetic combinations.

Comparison of mitosis and meiosis

Mitosis	Meiosis
One division resulting in two daughter cells	Two divisions resulting in four daughter cells
Number of chromosomes is unchanged	Number of chromosomes is halved
Homologous chromosomes do not associate in pairs	Homologous chromosomes pair up
Crossing over does not occur	Crossing over occurs and chiasmata form
Daughter cells are genetically identical	Daughter cells are genetically different

Meiosis and variation

In the long term, if a species is to survive in a constantly changing environment and to colonise new environments, sources of variation are essential. There are three ways of creating variety:

- During sexual reproduction the genotype of one parent is mixed with that of the other when haploid gametes fuse.
- Independent assortment results in gametes containing different combinations of chromosomes.
- During crossing over equivalent parts of homologous chromosomes may be exchanged thus producing new combinations and the separation of linked genes.

quickfire

㊴ State three differences between meiosis and mitosis.

quickfire

㊵ Explain how meiosis can give rise to genetically variable gametes.

Summary: Basic Biochemistry and Cell Structure

Basic biochemistry

Chemical properties of carbohydrates, fats and proteins related to chemical structure

Carbohydrates

A source of energy, polymers add strength and support

Lipids

Energy stores, insulation and protection, and component of cell membrane

Proteins

Enzymes, hormones, antibodies, transport and structural

Water is an important solvent and is involved in biochemical reactions

Cell structure

Prokaryotes

Simple organisms such as bacteria with no membrane-bound organelles

Eukaryotes

- Plants, animals, fungi and protoctists with membrane-bound organelles
- Variety of organelles

Cell membrane

- Consists of phospholipids and proteins
- Fluid Mosaic model
- Transport of materials by diffusion, facilitated diffusion, active transport and osmosis

Enzymes

- Globular proteins
- Properties related to tertiary structure
- Combine with a substrate to form an enzyme-substrate complex
- Affected by factors such as temperature, pH, and the concentration of the reactants
- Inhibited by competitive and non-competitive inhibitors
- Used widely in industry in immobilised form

Cell division

Mitosis

- Asexual reproduction and growth and repair of cells
- Daughter cells genetically identical to the parent

Meiosis

- Haploid gametes produced
- Daughter cells genetically different

DNA

- Double helix with base pairs bonded together
- Deoxyribose sugar
- Sequence of bases called the genetic code
- Replicates during cell division
- RNA differs from DNA

BY2: Biodiversity and Physiology of Body Systems

This unit gives an overview of the variety of organisms with an emphasis on comparative adaptations. Evolution has brought about the existence of the biodiversity of life forms on Earth. Taxonomy is the scientific study of the diversity of living organisms. DNA technology can be used to investigate how closely related organisms are. Evidence suggests that life evolved in an aquatic environment, and over long periods of time some groups of organisms adapted to a terrestrial existence. An increase in size meant that gaseous exchange by diffusion was inadequate, and larger and more advanced multicellular animals developed specialised gaseous exchange surfaces. These organisms also needed a rapid means of transport of materials provided by an internal transport system with a heart to pump blood through blood vessels. Angiosperms developed two distinct systems of tubes to transport water and sugars. Plants and animals also evolved reproductive adaptations to enable a successful terrestrial existence. Plants and animals differ markedly in their methods of nutrition; plants are autotrophic whereas animals are heterotrophic. Animals are ultimately dependent on plants for their nutrition, either directly or indirectly, and have adapted to different diets.

Revision checklist

Tick column 1 when you have completed brief revision notes.
Tick column 2 when you think you have a good grasp of the topic.
Tick column 3 during final revision when you feel you have mastery of the topic.

		1	2	3	Notes
	Biodiversity and evolution				
p41	Evolution				
p42	Classification				
p43	The five kingdom classification				
p44	Selected animal phyla				
p45	Chordata				
p46	Evidence of common ancestry				
p47	Using genetic evidence				
	Adaptations for gaseous exchange				
p48	Problems associated with increase in size				
p50	Gas exchange in fish				
p52	Adaptations of vertebrate groups to gaseous exchange on land				
p53	Gas exchange in insects				
p54	The structure of the human respiratory system				
p55	Gaseous exchange in plants				
p56	Mechanisms of opening and closing of stomata				

Revision checklist

		1	2	3	Notes
	Transport in animals and plants				
p57	Open and closed systems				
p58	Single and double circulations				
p58	Structure and function of blood vessels				
p60	The heart-cardiac cycle				
p61	Control of heartbeat				
p62	Blood				
p62	The transport of oxygen				
p65	Chloride shift – the transport of carbon dioxide				
p66	Intercellular fluid				
p67	Transport in plants				
p70	Transpiration				
p71	Mesophytes, xerophytes and hydrophytes				
p73	Translocation				
	Reproductive strategies				
p75	Asexual and sexual reproduction				
p76	Gamete production and fertilisation				
p77	Adaptations of animals to life on land				
p78	Colonisation of land by Angiosperms				
	Adaptations for nutrition				
p79	Methods of nutrition				
p81	Processing food in the digestive system				
p82	The structure of the human gut				
p83	Digestion				
p85	Absorption				
p86	Adaptations to different diets				
p87	Ruminants				
p88	Parasitic nutrition				

Biodiversity = a measure of the number of species on the planet.

Extinction = the loss of species.

Species diversity = the number of different species and the proportion of each species within a given area or community.

Grade boost

Research an ecosystem currently under threat due to human activity.
For a named endangered species in that ecosystem list the factors that have led to its endangered status. Consider the steps being made to conserve the species.

quickfire

① State the two main causes for the decline in the numbers of tigers.

》 Pointer

Evolutionary history shows that biodiversity has gone through several bottlenecks, called mass extinctions, followed by radiations of new species.

Biodiversity and evolution

Over the last 200 years certain human activities have had negative effects on the environment and this has, in turn, affected the survival of plants and animals. Now scientists have come to realise that there is a **biodiversity** crisis, a rapid decrease in the variety of life on Earth.

Human activities are altering ecosystems upon which they and other species depend. Tropical rain forests are being destroyed at an alarming rate to make room for, and to support, the increase in the human population. In the oceans, stocks of many fishes are being depleted by over-harvesting, and some of the most productive and diverse areas, such as coral reefs and estuaries, are being severely stressed. Globally, the rate of species loss may be as much as 50 times higher than at any time in the past 100,000 years. Human alteration to habitat is the single greatest threat to biodiversity on the planet.

Extinction is a natural process that has been taking place since life first evolved. It is the current *rate* of extinction that underlies the biodiversity crisis. Scientists believe that the normal 'background' rate of extinction is one out of every million species per year. It is now estimated that human activity in tropical areas alone has increased extinction rates between 1000 and 10,000 times! Massive destruction of habitats throughout the world has been brought about by agriculture, urban development, forestry, mining, and environmental pollution. Marine life has also been affected. About one-third of the planet's marine fish species rely on coral reefs. At the current rate of destruction about half of the reefs could be lost in the next 20 years.

The vast majority of Earth's earlier occupants, including the large and once dominant dinosaurs and tree ferns, have become extinct largely as a result of climatic, geological and biotic changes. At the present time, human activity has taken over as the main cause of species extinction. The main causes for the decline in numbers of larger mammals such as mountain gorillas, giant pandas, tigers and polar bears are loss of habitat; over-hunting by humans: competition from introduced species. Other species are also threatened by additional causes such as deforestation, pollution and drainage of wetlands.

It is now recognised that each species may represent an important human asset, a potential source of food, useful chemicals, or disease-resistant genes. For example, of the many plants growing in the tropical rain forests there may be some with medicinal properties. The extinction of any plant species before their chemical properties have been investigated could amount to an incalculable loss. There is therefore a need for species conservation, the planned preservation of wildlife.

Evolution

The term is **evolution** is used specifically for the processes that have transformed life on Earth from its early beginnings to the vast diversity of fossilised and living forms that are known today.

The theory of evolution was first put forward by Charles Darwin. During his visit to South america and the Galapagos Islands Darwin accumulated geological and **fossil** evidence that supported the idea that life changes with time. In 1859 he proposed natural selection as the force that causes changes in populations.

Darwin studied the fourteen different species of finches found on the Galapagos Islands. In geological terms the islands were recently formed and any animals found there must have reached the islands from the mainland, some 600 miles away. Finches are unable to fly long distances and Darwin suggested that one ancestral species of finch had reached the islands with the help of the prevailing winds. As there were no other bird species inhabiting the islands, there was a variety of food available to the colonising finches. He noticed how individual finches differed from one island to the next. The main differences were in the size and shape of their beaks and these were related to the different type of food eaten, for example insects, seeds, fruit.

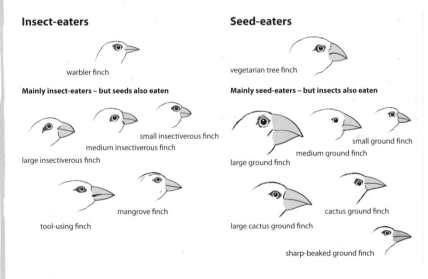

Insect-eaters

warbler finch

Mainly insect-eaters – but seeds also eaten

small insectiverous finch
medium insectiverous finch
large insectiverous finch

mangrove finch

tool-using finch

Seed-eaters

vegetarian tree finch

Mainly seed-eaters – but insects also eaten

small ground finch
medium ground finch
large ground finch

cactus ground finch
large cactus ground finch

sharp-beaked ground finch

Different types of finch beaks.

It seemed that on each island the characteristics that best suited a particular finch to its environment were inherited by the offspring. Darwin suggested that the finches had developed from a common ancestor and that the type of beak had developed over time and become specialised to feed on a particular food source. This is an example of adaptive radiation.

Grade boost

You should know about the principles of modern classification that show how organisms may be related through evolution by the number of common characteristics they share.

Grade boost

Be prepared to place taxa in the correct order.

② What is the definition of a species?

Grade boost

Note that the genus name is given a capital letter and the species name a lower case. Both are in italics.

③ Why is it important for scientists to use Latin names for organisms mentioned in research papers?

Classification

As there is a vast number of species it makes sense to organise them into manageable groups. The grouping of organisms is called classification.

When describing plants and animals, **taxonomists** look for differences and similarities between them and place similar organisms closely together and dissimilar ones further apart. A classification system based on large groups being divided up into progressively smaller groups is said to be hierarchical.

The natural classification used today was devised by the Swedish scientist, Linnaeus in the 18th century. In this scheme, organisms are grouped together according to their basic similarities. A hierarchical system has been devised to distinguish large groups of organisms with a series of rank names to identify the different levels within the hierarchy. A **taxon** is a level in this classification hierarchy and is a collection of organisms sharing some basic features.

- Kingdom – the largest taxonomic grouping, e.g. animals, plants.
- Phylum – a large grouping of all the classes that share some common features, e.g. Arthropods (includes insects, spiders, centipedes and millipedes, crustaceans).
- Class – a grouping of similar orders, e.g. Insecta (insects).
- Order – grouping of related families, e.g. Orthoptera (includes locusts, grasshoppers and crickets).
- Family – a grouping of similar genera (plural of genus), e.g. Rosaceae (rose).
- Genus – a group of species that are very closely related, for example Locusta.
- Species – a group of organisms which share a large number of common characteristics and which can interbreed to produce fertile offspring, for example *Locusta migratoria*.

Binomial system

At one time, new organisms were given different names in different countries with many having common names which differed from one part of a country to another! This led to confusion among scientists if a particular species needed to be named and described in a scientific research paper.

In 1753 Linnaeus overcame this problem by devising a common system of naming organisms. Each organism is given two names, the generic name (genus) and the specific name (species). The system is called the binomial system and is based on using Latin as an international language.

The five kingdom classification

Living organisms are divided into five large groups or kingdoms.

Prokaryotae

Unicellular organisms and include bacteria and blue-green algae. They have no internal cell membranes, no nuclear membrane, no endoplasmic reticulum, no mitochondria and no Golgi body. They possess a cell wall but it is not made of cellulose.

Protoctista

Mostly small eukaryotic organisms, with membrane-bound organelles and a nucleus with a nuclear membrane. In this kingdom are found the organisms that are neither plants nor animals nor fungi. The kingdom includes algae, the water moulds, the slime moulds and the protozoa.

Fungi

Eukaryotic, the body consisting of a network of threads called hyphae, forming a mycelium. There is a rigid cell wall made of chitin. They do not have photosynthetic pigments and feeding is heterotrophic; all members of the group are either saprophytic or parasitic. In some sub groups, the hyphae have no cross-walls, but in others cross-walls, or septa, are present. Reproduction is by spores that lack flagella. Examples are *Penicillium*, yeast, mushroom.

Plants

Multicellular and carry out photosynthesis. The cells are eukaryotic, have cellulose walls, vacuoles containing cell sap, and chloroplasts containing photosynthetic pigments. The main plant phyla include mosses and liverworts, ferns, conifers and flowering plants.

Animals

Multicellular, heterotrophic, eukaryotes with cells lacking a cell wall, and show nervous coordination.

The animal kingdom is divided into two main groups:

- Non-chordates, often called invertebrates. Examples include segmented worms, molluscs and arthropods.
- Chordates – all but the simplest of the chordates have a vertebral column and are therefore referred to as vertebrates and include fish, amphibians, reptiles, birds and mammals.

Grade boost
You should learn the basic features which distinguish the five kingdoms.

(4) Some fungi are plant-like in appearance. Why are they placed in a separate kingdom?

Pointer
The flowering plants (Angiosperms) are the most dominant plant group on Earth. They include all our major crops and are therefore an important food source. Their flowers have seeds that are enclosed in a fruit formed from the ovary wall.

Pointer
95% of all animals are invertebrates and only 5% are vertebrates.

≫ Pointer

It is an incredible fact that 75% of all animal species are insects!

≫ Pointer

The one main disadvantage of the exoskeleton is that it is fixed in size and does not grow with the animal. This contrasts with the vertebrate endoskeleton which increases in size as the body grows. In order to grow, an arthropod must periodically shed its exoskeleton (ecdysis). This leaves the animal especially vulnerable as the new exoskeleton hardens.

⑤ State two functions of an exoskeleton.

Selected animal phyla

Annelids

There are 8,000 named species of annelids. They include earthworms, leeches and lugworms. All members of the phylum have the following common features:

- A long, thin segmented body, the segments being visible externally as rings with a body divided internally by partitions (septa).
- A fluid-filled body cavity (haemocoel).
- A hydrostatic skeleton.
- A head end with a primitive brain and a nervous system running the length of the body.
- A thin permeable skin, through which gaseous exchange occurs.
- A closed circulatory system containing an oxygen-carrying pigment.

Arthropods

The Arthropoda is divided into four classes:

- *Myriopoda* – they have many pairs of legs, one or two per segment, examples are millipedes and centipedes.
- *Crustacea* – have between 10 and 20 pairs of legs, e.g. crab.
- *Spiders* – have four pairs of legs.
- *Insects* – have three pairs of legs.

The arthropods have the following common features: a body divided into segments, a body further divided into head, thorax and abdomen, a well-developed brain, a hard outer exoskeleton, paired jointed legs, an open circulatory system and a cavity which surrounds the body organs.

Two important evolutionary developments of this phylum are:

- *Jointed legs* modified to perform a variety of functions, including walking, swimming, jumping, feeding, reproduction and, where present, ventilation of the gills.
- *Exoskeleton* – the outermost layer of cells of the body secretes a thick cuticle, which consists mainly of chitin. This performs several functions:
  - Protection of internal organs, protection from predators
  - A point of attachment for muscles
  - Support – for small animals a hollow tubular structure surrounding the body provides greater support than a solid cylindrical rod within it (an endoskeleton as in vertebrates) made from the same quantity of material
  - In most terrestrial arthropods the exoskeleton is covered with a layer of wax which reduces water loss.

Chordata

There are 60,000 named species of Chordates. Examples include frogs, snakes, eagles and humans.

Vertebrates possess:

- a vertebral column or backbone
- a well-developed brain, enclosed in a cranium.

The vertebrates are subdivided into five classes:

- Fish – aquatic forms with scales, fins and gills.
- Amphibians – these were the first land vertebrates, partly terrestrial and partly aquatic. They have a soft, moist skin. The eggs are fertilised externally in water where they also develop. Young (larvae) are aquatic and have gills; adults are usually terrestrial and have simple lungs.
- Reptiles – mainly terrestrial and have a dry skin with scales. They have lungs. The eggs are fertilised internally, covered with a shell and laid on land.
- Birds – similar to reptiles in many ways. Differences are mainly due to the ability to fly and the development of feathers, with fore-legs developed as wings. They have lungs. Eggs have hard shells.
- Mammals – skin with hair. Young are born alive and are fed on milk. Have lungs. They are further subdivided into two groups:
  - Marsupials, e.g. kangaroo – young are born at a very immature state and develop in female's pouch.
  - Placentals – young undergo considerable development in the mother's womb, and receive nourishment via the placenta before they are born.

The phylogenetic relationships of different species can be represented by a diagram called a phylogenetic tree. The closer the branches, the closer the evolutionary relationship.

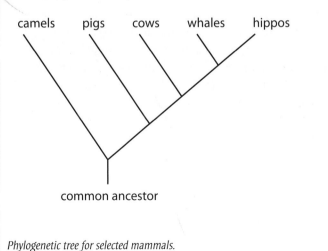

Phylogenetic tree for selected mammals.

Key Terms

Chordata = the scientific name for vertebrates.

Phylogeny = the evolutionary relationship between organisms.

Grade boost

Note the progression from water to land in the vertebrate classes. Different terrestrial groups have adapted to life on land in different ways.

Grade boost

You should have an overview of a variety of organisms and their comparative adaptations. You are not expected to memorise the detailed classification of any of the groups.

quickfire

⑥ Which are most closely related, hippos and whales or hippos and cows?

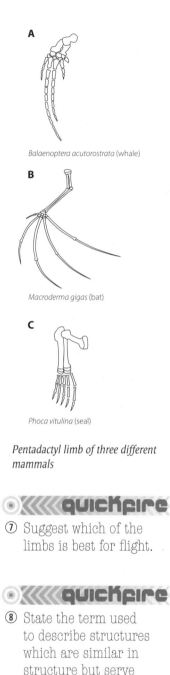

A

Balaenoptera acutorostrata (whale)

B

Macroderma gigas (bat)

C

Phoca vitulina (seal)

Pentadactyl limb of three different mammals

quickfire

⑦ Suggest which of the limbs is best for flight.

quickfire

⑧ State the term used to describe structures which are similar in structure but serve different functions.

Evidence of common ancestry

The theory of evolution suggests that widely separated groups of organisms share a common ancestor. Therefore it would be expected that they share certain basic structural features. How similar they are should indicate how closely related they are in terms of evolution. Groups with little in common are assumed to have diverged from a common ancestor much earlier in geological history than groups which have a lot in common.

Using physical features

In deciding how closely related two organisms are a biologist needs to look for structures that, though they may serve quite different functions, are similar in structure, suggesting a common origin. Such structures are said to be homologous. A good example of this is the pentadactyl limb of Chordata (vertebrates).

A pentadactyl limb has five digits. It is found in the four classes of terrestrial vertebrates: amphibians, reptiles, birds and mammals. The structure of the limb is basically the same in all the classes. However, the limbs of the different vertebrates have become adapted for different functions, such as grasping, walking, swimming and flying, in a selection of vertebrates.

Examples of the pentadactyl limb modified for different functions are: the human arm, the wing of a bat, the flipper of a whale, the wing of a bird, the leg of a horse.

Using information such as this it is possible to construct evolutionary trees where the end products of evolution have certain structural features in common with each other and with the ancestral stock from which they arise. The more similar two organisms are, the more recently they are assumed to have diverged.

However, there is a possible danger in assuming that just because two animals look similar that they are related. Consider a shark, a porpoise and a penguin. One is a fish, one is a bird and the other is a mammal. By studying the skeleton of the fore-limbs it is possible to deduce that the dolphin and the penguin have modified pentadactyl limbs, but the fish does not. The shark and dolphin are animals that have similar fins because they live in similar environments and have become adapted to that environment, not because they have a common ancestor. That is, the structures, limbs, are performing the same function. Such structures are described as analogous. Another example is the wings of birds and insects.

Using genetic evidence

During the course of evolution when one species gives rise to another the new species will have some differences in the sequence of nucleotide bases in the DNA. Over time the new species will accumulate more differences in the DNA. Therefore one would expect species that are more closely related to show more similarity in their DNA base sequences than species that are more distantly related.

DNA analysis has been used to confirm evolutionary relationships and can reduce the mistakes made in classification due to **convergent evolution**.

- The technique of DNA hybridisation involves the extraction and comparison of the DNA of two species. The sequence of bases is compared and the more alike the sequences, the closer the organisms are related in terms of evolution.

- The sequence of amino acids in proteins is determined by DNA. Therefore the degree of similarity in the amino acid sequence of the same protein in two species will reflect how closely related the two species are. Part of the fibrinogen molecule of various mammals has been compared and the sequence has been found to differ in varying degrees from one species to another and this has enabled scientists to draw up a possible evolutionary tree for mammals.

- The proteins of different species can also be compared using immunological techniques. The principle behind this involves the fact that antibodies of one species will respond to specific antigens on proteins, such as albumin, in the blood serum of another. When antibodies respond to corresponding antigens a precipitate is formed. The greater the degree of precipitation the closer the evolutionary relationship.

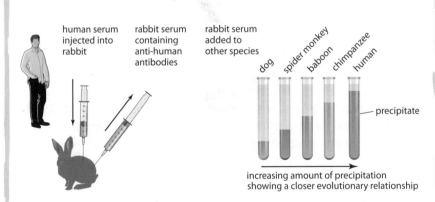

Immunological comparisons of human serum with that of other species.

Key Terms

Mutations = a permanent change in the DNA or chromosome number of the cell.

Convergent evolution = the tendency of unrelated organisms to acquire similar structures.

≫ Pointer

To work out how closely related two species of primates are, such as humans and chimpanzees, the DNA strands from both species are extracted, separated and cut into fragments. The fragments from the two species are then mixed and analysed. This technique gives results which show that chimpanzees and humans have 97.6% DNA in common, whereas humans and rhesus monkeys have 91.1% DNA in common. Recent studies using this technique have also shown that the hippopotamus and whale are closely related.

⩕ Grade boost

Comparing the DNA and proteins of different species helps scientists to determine the evolutionary relationships between them.

Grade boost

All living organisms exchange gases with the environment.

An aquatic environment is fairly constant but life on land can be more extreme. In order to survive, living organisms have adapted in different ways.

≫ Pointer

Gas exchange is the process by which oxygen reaches cells and carbon dioxide is removed from them. It should not be confused with respiration.

quickfire

⑨ Name three factors that affect the rate of diffusion of substances into cells.

Adaptations for gaseous exchange

Problems associated with increase in size

Animals and plants have evolved special gas exchange surfaces so that diffusion of gases into and out of cells can take place rapidly and efficiently. The gills of a fish, the alveoli in the lungs of a mammal and the spongy mesophyll cells in the leaves of a plant, are all excellent gas exchange surfaces. In order to achieve the maximum rate of diffusion, a respiratory surface must:

- Have a sufficiently large surface area relative to the volume of the organism to satisfy the needs of the organism.
- Be thin, so that diffusion paths are short.
- Be permeable to allow the respiratory gases through.
- Be moist to allow a medium in which gases can dissolve.

In simple single-celled organisms such as *Amoeba*, the cell membrane is the gas exchange surface. The organism lives in water and the diffusion of gases occurs over the whole of the body surface. A single cell has a large surface area compared to its volume. It is said to have a large surface area to volume ratio. The efficiency of gaseous diffusion satisfies the needs of the organism. The membrane is thin, moist, and the diffusion paths are short.

There must be a limit to cell size, and a point is reached where the diffusion path is so long that the process of diffusion becomes inefficient. In terms of evolution, the only way that organisms could become larger was to aggregate cells together, that is, to become multicellular. However, the larger the organism, the smaller will be the surface area to volume ratio. Also, materials need to be exchanged between different organs as well as between the organs and the environment. This means that the gaseous requirements provided by diffusion through the cell surface are insufficient to meet the needs of the organism. In effect, the process of diffusion is too slow.

- Simple multicellular animals, such as the earthworm, have modest oxygen requirements because they are slow moving and so have a very low metabolic rate. Oxygen and carbon dioxide diffuse across the skin surface and they do not have any special gas exchange organs.

- Oxygen diffuses into the blood capillaries beneath the skin surface and is carried in vessels to the cells with carbon dioxide being transported in the opposite direction. In this way the blood system maintains a diffusion gradient at the respiratory surface. However, once oxygen is inside the body it needs to be transported some distance to the many internal cells. The blood also contains a respiratory pigment for the transport of oxygen.

- Flatworms are aquatic animals that have evolved a flattened shape. This considerably increases the surface area to volume ratio and ensures that no part of the body is far from the surface, i.e. diffusion paths are short.

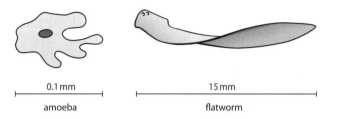

0.1 mm   15 mm

amoeba   flatworm

- Larger and more advanced multicellular animals, such as insects, fish, reptiles and mammals, have a high metabolic rate, i.e. they need more energy and have a high requirement for oxygen. Also, the larger the organism, the smaller is the surface area to volume ratio. With an increase in size and specialisation, tissues and organs became more dependent on one another. In order to make gas exchange more efficient, these organisms have developed a specialised exchange surface to compensate for the increased oxygen demand:
  - In aquatic insects and fish the exchange surfaces for respiratory gases take the form of gills.
  - Terrestrial animal groups such as birds, reptiles and mammals have developed lungs.

All of these different mechanisms need a means of ventilation to supply the respiratory surfaces with a fresh supply of oxygen and to maintain diffusion gradients. That is, the function of a ventilation mechanism is to move the respiratory medium, air or water, over the respiratory surface.

These different animal groups have also developed:

- An internal transport system – provided by a blood circulation system to move gases between respiring cells and the respiratory surface.
- A respiratory pigment in the blood – to increase its oxygen-carrying capacity.

Gas exchange in fish

⑫ Name three structural features of fish gills which make them efficient gaseous exchange organs.

Aquatic organisms have a problem with gaseous exchange because water contains far less oxygen than air and the rate of diffusion in water is slower. Also, water is a denser medium than air and does not flow as freely. As fish are very active, they need a good supply of oxygen. In fish, gaseous exchange takes place across a special surface, the gill, over which a one-way current of water is kept flowing by a specialised pumping mechanism. The density of the water prevents the gills from collapsing and lying on top of each other, which would reduce the surface area. Gills are made up of many folds, providing a large surface area over which water can flow, and gases can be exchanged.

Fish are divided into two main groups according to the material that makes up their skeleton:

- Cartilaginous fish, e.g. sharks, have a skeleton made entirely of cartilage. Nearly all live in the sea. Just behind the head on each side are five gill clefts which open at gill slits. Water is taken into the mouth and is forced through the gill slits when the floor of the mouth is raised. Blood travels through the gill capillaries in the same direction as the sea-water. Gas exchange in such a parallel flow is relatively inefficient.

- Bony fish have an internal skeleton made of bone and the gills are covered with a flap called the operculum. Bony fish inhabit both fresh and sea water and are by far the most numerous of aquatic vertebrates. Gas exchange involves a counter-current flow arrangement whereby blood in the gill capillaries flows in the opposite direction to the water flowing over the gill surface.

In bony fish there are four pairs of gills in the pharynx and each gill is supported by a gill arch. Along each gill arch are many pairs of gill filaments and on these are the gas exchange surfaces, the gill lamellae. Out of water the gill collapses as the gill filaments lie on top of each other and stick together. However, the gill filaments are supported in water with the gill lamellae providing a large surface area. The gill lamellae contain blood capillaries and these take up oxygen from the water and carbon dioxide passes out.

Grade boost

Be prepared to draw arrows on given diagrams to indicate the direction of blood flow and water flow.

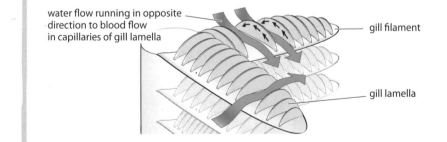

water flow running in opposite direction to blood flow in capillaries of gill lamella

gill filament

gill lamella

Arrangement of fish gills and direction of water flow.

Gills provide:

- A specialised area rather than using the whole body surface.
- A large surface extended by the gill filaments.
- An extensive network of blood capillaries to allow efficient diffusion and haemoglobin for oxygen carriage.

To increase efficiency, water needs to be forced over the gill filaments by pressure differences so maintaining a continuous, unidirectional flow of water. A lower pressure is maintained in the opercular cavity than in the bucco-pharynx. The operculum acts as both a valve, permitting water out, and as a pump, drawing water past the gill filaments. The mouth also acts as a pump.

The ventilation mechanism for forcing water over the gill filaments operates as follows:

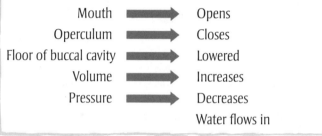

Mouth ➡	Opens
Operculum ➡	Closes
Floor of buccal cavity ➡	Lowered
Volume ➡	Increases
Pressure ➡	Decreases
	Water flows in

Counter-current flow

The orientation of the gas-exchange surfaces is such that as the water passes from the pharynx into the opercular chamber, it flows between the gill lamellae in the opposite direction to the blood flow. This increases efficiency because the diffusion gradient between the adjacent flows is maintained over the whole length of the gill filament. That is, the blood always meets water with a relatively higher oxygen content. This system allows the gills of a bony fish to remove 80% of the oxygen from water. This is three times the rate of extraction of oxygen from air in human lungs. This high level of extraction is essential to fish as there is around 25 times less oxygen in water, compared to air.

The counter-current system maintains a diffusion gradient along the whole length of the gill plate.

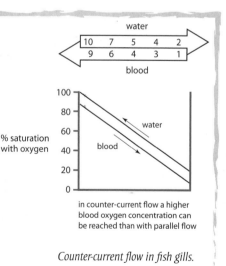

Counter-current flow in fish gills.

⑬ How does counter-current flow improve the efficiency of oxygen uptake?

Adaptations of vertebrate groups to gaseous exchange on land

The problem for all terrestrial organisms is that water evaporates from the body surface resulting in dehydration. Gas exchange surfaces need to be thin, permeable surfaces with a large area. These features conflict with the need to conserve water. Vertebrates and insects have evolved different methods of overcoming this problem.

Life is thought to have evolved in water with animals adapting in order to colonise the land and some becoming highly modified for flight. Gills do not function out of water and there was a need for vertebrates to evolve a different form of gaseous exchange surface, the lung. Birds and mammals are particularly active and adapted for exchange with air, a less dense medium, instead of water, so have internal lungs to minimise loss of water and heat.

Amphibians

The amphibians include frogs, toads and newts. This group was probably the first vertebrate group to colonise the land. Frogs are typical of amphibians in that they live in moist habitats as they require water for fertilisation. The larvae (tadpoles) also live in water and have gills. The transition from larva to land living adult involves great changes in the body form, known as metamorphosis. The inactive adult uses the moist skin as a respiratory surface and this provides sufficient oxygen for its needs. However, when active, as in mating, the frog uses lungs as a respiratory surface.

Reptiles

Reptiles include crocodiles, lizards, and snakes. The few present-day reptiles are descendants of a once very successful group of animals, including the dinosaurs, which dominated the Earth about 200 million years ago. They are far better suited to life on land than amphibians. Reptiles can move on all four limbs without the trunk of the body touching the ground. Pairs of ribs project from the vertebrae (backbone). Ribs provide support and protection to the organs in the body cavity. Ribs are also involved in the ventilation of the lungs. The lung also has a more complex internal structure than that of amphibians with the in-growth of tissues increasing the surface area for gas exchange.

Birds

The lungs of birds have an internal structure similar to that of mammals. However, large volumes of oxygen are needed to provide the energy for flight. Ventilation of the lungs in birds is far more efficient than in other vertebrates and is assisted by a system of air sacs, which function as bellows. Ventilation of the lungs is brought about by the movement of the ribs. During flight the action of the flight muscles ventilates the lungs.

Gas exchange in insects

To reduce water loss, terrestrial organisms need to have waterproof coverings over their body surfaces. To overcome this problem insects have evolved a rigid exoskeleton which is covered by a cuticle. Insects have a relatively small surface area to volume ratio and so cannot use their body surface to exchange gases by diffusion.

They have evolved a different system of gaseous exchange to other land animals. Gas exchange occurs through paired holes, called spiracles, running along the side of the body. The spiracles lead into a system of branched chitin lined air-tubes called tracheae. The spiracles can open and close like valves. This allows gaseous exchange to take place and also reduces water loss.

Resting insects rely on diffusion to take in oxygen and to remove carbon dioxide. During periods of activity, such as during flight, movements of the abdomen ventilate the tracheae. The ends of the tracheal branches are called tracheoles; here gaseous exchange takes place resulting in oxygen passing directly into the cells.

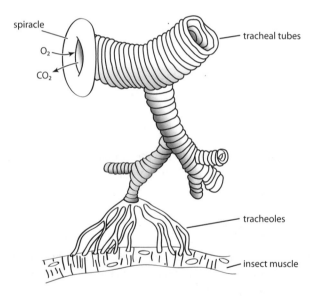

Part of an insect tracheal system.

>> *Pointer*

Most insects possess wings and are extremely efficient in the air. Flight involves a high expenditure of energy and so insects need a good supply of oxygen.

>> *Pointer*

Every cell of an insect is only a very short distance from tracheoles and therefore the diffusion pathway is short.

 quickfire

⑭ How do insects prevent excessive water loss from their tracheal system?

quickfire

⑮ Suggest why the tracheal system limits the size of insects.

quickfire

⑯ State two advantages of using a tracheal system for gas exchange.

Grade boost

Be prepared to label a given diagram of the respiratory system.

quickfire

(17) List in the correct order the structures that air passes through from the gas exchange surface to the atmosphere.

quickfire

(18) What is the function of cartilage in the trachea?

Grade boost

The essential features of exchange surfaces are the same in all organisms.

Lungs supply a large surface area, increased by alveoli, lined with moisture for the dissolving of gases, thin walls to shorten the diffusion path and an extensive capillary network for rapid diffusion and transport, to maintain diffusion gradients.

⟫ Pointer

The diffusion pathway is short because the alveoli have only a single layer of epithelial cells and the blood capillaries also have only a single layer of cells.

The structure of the human respiratory system

The lungs are enclosed within an airtight compartment, the thorax, at the base of which is a dome-shaped sheet of muscle called the diaphragm. Air is drawn into the lungs via the trachea. The lungs consist of a branching network of tubes called bronchioles arising from a pair of bronchi.

Gas exchange in the alveolus

The gas exchange surfaces are the alveoli (air sacs) which provide a very large surface area relative to the volume of the body. They are well suited as a gas exchange surface because the walls are thin, providing a short diffusion path. Each alveolus is covered by an extensive capillary network to maintain diffusion gradients, because blood is always taking oxygen away from the alveolus and returning with carbon dioxide.

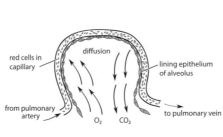

Diffusion of gases in an alveolus.

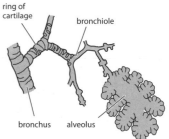

Part of human respiratory system.

Deoxygenated blood enters the capillaries around the alveolus. Oxygen diffuses out of the alveolus into the blood in the capillary. Carbon dioxide diffuses out of the capillary into the air in the alveolus.

Ventilation of the lungs

Mammals ventilate their lungs by negative pressure breathing, forcing air down into the lungs. That is, if air is to enter the lungs then the pressure inside them must be lower than atmospheric pressure.

	Inspiration	Expiration
External intercostal muscle	Contracts	Relaxes
Ribs	Up and out	Down and in
Diaphragm	Contracts and flattens	Relaxes
Volume of thorax	Increases	Decreases
Pressure in thorax	Decreases	Increases
Atmospheric pressure	Greater, so air moves in	Less, so air moves out

Gaseous exchange in plants

The leaf as an organ of gaseous exchange

To enable gaseous exchange to take place efficiently:

- The leaf blade is thin and flat with a large surface area.
- The spongy mesophyll tissue allows for the circulation of gases.
- The plant tissues are permeated by air spaces.
- The stomatal pores permit gas exchange.

Gases diffuse through the stomata along a concentration gradient. Once inside the leaf the gases in the sub-stomatal air chambers diffuse through the intercellular spaces between the spongy mesophyll cells and into the cells. The direction of diffusion depends on the environmental conditions and the requirements of the plants. It is the net exchange of carbon dioxide and oxygen in relation to respiration and photosynthesis that matters.

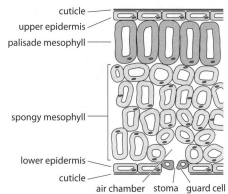

cuticle
upper epidermis
palisade mesophyll
spongy mesophyll
lower epidermis
cuticle
air chamber   stoma   guard cell

T.S. section through a leaf.

>> *Pointer*
There is a short, fast diffusion pathway in plants.

quickfire

⑲ State the functions of the palisade and spongy mesophyll layers.

Grade boost

As the rate of photosynthesis is greater than the rate of respiration, during the day the overall gas released is oxygen. At night only respiration occurs, so the gas released is carbon dioxide.

quickfire

⑳ State two differences between gas exchange in a plant leaf and gas exchange in a terrestrial insect.

Adaptations of the leaf for photosynthesis

To ensure the efficient absorption of light, the leaf shows the following adaptations:

- Leaves have a large surface area to capture as much sunlight as possible.
- Leaves can orientate themselves so that they are held at an angle perpendicular to the sun during the day to expose the maximum area to the light.
- Leaves are thin to allow light to penetrate the lower layers of cells.
- The cuticle and epidermis are transparent to allow light to penetrate to the mesophyll.
- Palisade cells are elongated and densely arranged in a layer, or layers.
- The palisade cells are packed with chloroplasts and arranged with their long axes perpendicular to the surface.
- The chloroplasts can rotate and move within the mesophyll cells. This allows them to arrange themselves into the best positions for the efficient absorption of light.
- The intercellular air spaces in the spongy mesophyll allow carbon dioxide to diffuse to the cells and oxygen can diffuse away.

>> *Pointer*

By opening and closing, the stomata control the quantity of gases diffusing into and out of the leaf. They also control the quantity of water vapour evaporating.

Grade boost

The mechanism of opening and closing of stomata is a favourite for examiners! It will help to review earlier work on osmosis. .

>> *Pointer*

Guard cells can change shape to open and close the stomata. Cells change shape because of changes in turgor.

Mechanism of opening and closing of stomata

Stomata are small pores found on the lower surface of a leaf. Each pore is bounded by two guard cells. Guard cells are unusual in having chloroplasts and unevenly thickened walls, with the inner wall being thick and the outer wall thin. The stomata allow exchange of gases between the atmosphere and the internal tissues of the leaf. However, water also evaporates from a plant though the stomata. This process is called transpiration. Plants wilt if they lose too much water. Since light hits the upper surface of the leaf, confining stomata to the lower surface reduces water loss. The presence of a waxy cuticle on the upper surface also reduces water loss significantly. In most plants the stomata close at night thus preventing the plant from needlessly losing water when the light intensity is insufficient for photosynthesis to take place.

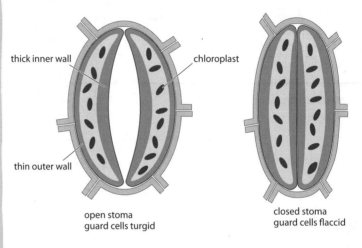

thick inner wall

chloroplast

thin outer wall

open stoma
guard cells turgid

closed stoma
guard cells flaccid

Surface view of stomata open and closed.

During the day the mechanism for stomatal opening occurs as follows:

- The chloroplasts in the guard cells photosynthesise, producing ATP.
- Using ATP potassium ion (K^+) pumps in the cell membranes of the surrounding epidermal cells actively transporting K^+ ions into the guard cells.
- Stored starch is converted to malate.
- The water potential in the guard cells is lowered (becomes more negative) and water enters by osmosis.
- The guard cells become turgid and curve apart because their outer walls are thinner than the inner walls, so the pore widens.

The reverse process occurs at night and the pore closes.

Transport in animals and plants

Multicellular organisms require a transport system to take materials from cells to exchange surfaces and from exchange surfaces to cells. Materials also have to be transported between the exchange surface and the environment, as well as to different organs in the body.

A transport system has the following common features:

- A suitable medium (blood) in which to carry materials.
- A closed system of vessels that contains the blood and forms a branching network to distribute it to all parts of the body.
- A pump, such as the heart, for moving the blood within vessels.
- Valves to maintain the flow in one direction.
- A respiratory pigment (absent in insects) which increases the volume of oxygen that can be transported.

Open and closed systems

Insects have an open blood system whereby blood is pumped at relatively low pressure from one main long, dorsal (top) tube-shaped heart running the length of the body. The blood is pumped out of this heart into spaces, collectively called a haemocoel, within the body cavity. The blood bathes the tissues directly, exchange of materials takes place, and there is little control over the direction of circulation. Blood slowly returns to the heart. Here valves and waves of contraction of the muscle wall move the blood forward to the head region where the open circulation is started again. There is usually no respiratory pigment in the insect as the blood of an insect does not transport oxygen. Oxygen is transported directly to the tissues via the tracheae.

Mammals have a closed circulation system whereby the blood circulates in a continuous system of tubes, the blood vessels. Blood is pumped by a muscular heart at high pressure, resulting in a rapid flow rate. Organs are not in direct contact with the blood but are bathed by tissue fluid seeping out from thin-walled capillaries. The blood contains a blood pigment which carries oxygen.

Even though the earthworm is a relatively simple organism compared to a mammal, it has a closed circulation system. It has dorsal and ventral vessels running the length of the body and these are connected by five 'pseudohearts'. Blood moves through the vessels by the pumping action of the 'pseudohearts'.

Grade boost

The greater the metabolic rate, the greater the need for rapid transport of glucose and oxygen.

quickfire

㉒ Apart from vessels and valves, state three features, giving examples, of a transport system in animals.

quickfire

㉓ State three differences between the blood system of an insect and that of a mammal.

quickfire

㉔ What is the advantage of a double circulation system?

❯❯ Pointer

Arteries and veins transport materials whereas gaseous exchange takes place at the capillaries.

Single and double circulations

- Closed circulation systems are of two types, depending on whether the blood passes through the heart once or twice in each circulation of the body.
- Fish have a single circulation. The heart pumps deoxygenated blood to the gills, oxygenated blood is then carried to the tissues, from there deoxygenated blood returns to the heart. Blood goes once through the heart during each circuit of the body.
- Mammals have a double circulation:
  - The pulmonary circulation – the right side of the heart pumps deoxygenated blood to the lungs. Oxygenated blood then returns to the left side of the heart.
  - Systemic circulation – the left side of the heart pumps the oxygenated blood to the tissues. Deoxygenated blood then returns to the right side of the heart.
  - In each circuit the blood passes through the heart twice, once though the right side and once through the left side.

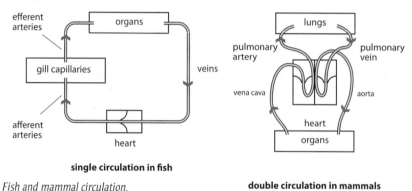

single circulation in fish

Fish and mammal circulation.

double circulation in mammals

Structure and function of blood vessels

There are three types of blood vessels: arteries, veins and capillaries.

Arteries and veins have the same basic three-layered structure, but the proportion of the different layers varies. In both arteries and veins:

- The innermost layer is the endothelium, which is one cell thick and provides a smooth lining to reduce friction and provide a minimum resistance to the flow of the blood.
- The middle layer is made up of elastic fibres and smooth muscle. This layer is thicker in the arteries than in the veins to accommodate changes in blood flow and pressure as blood is pumped from the heart.
- The outer layer is made up of collagen fibres which are resistant to over-stretching.

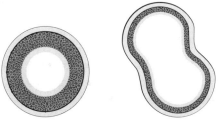

| artery | vein | capillary |

Diagrams of an artery, vein and capillary.

Arteries carry blood away from the heart. The arteries have thick, muscular walls to withstand the high pressure of blood received from the heart. The contraction of the arterial muscles also helps to maintain pressure as the blood is transported further from the heart.

The arteries branch into smaller vessels called arterioles that further subdivide into thin-walled capillaries. The capillaries form a vast network which penetrates all the tissues and organs of the body. Blood from the capillaries collects into venules, which in turn empty blood into veins, from which it is returned to the heart.

Veins have larger diameters and thinner walls than arteries as the pressure and flow is reduced. Veins have semi-lunar valves along their length to ensure flow in one direction (prevent back flow); these are not present in arteries apart from the aortic valves.

Capillaries are thin walled, consisting only of a layer of endothelium so their walls are permeable to water and dissolved substances such as glucose. It is at the capillaries that the exchange of materials between the blood and the tissues takes place.

Capillaries have a small diameter, and friction with the walls slows the blood flow. Although the diameter is small, there are many capillaries in the capillary bed providing a large total cross-sectional area, which further reduces blood flow. This low velocity in very thin-walled vessels enhances their ability to exchange materials with the surrounding tissue fluid.

Grade boost

Construct a table comparing an artery and a vein. In an exam question it is important to make a true comparison. It is insufficient to state that veins have a large lumen without adding that arteries have a small lumen.

Grade boost

Arteries carry blood away from the heart; veins carry blood to the heart.

≫ Pointer

It may at first seem that the blood should travel faster through capillaries than through arteries, because the diameters of the capillaries are much smaller. However, it is the total cross-sectional area delivering the blood that determines the flow rate.

quickfire

㉕ Give two reasons why arteries have a thick muscular wall.

quickfire

㉖ The capillaries slow down the flow of blood at the tissues. What is the significance of this?

Systole = the contraction phase of the cardiac cycle.

Diastole = the relaxation phase.

Myogenic = the heartbeat is initiated from within the muscle itself and is not due to nervous stimulation.

≫ *Pointer*

A pump to circulate blood is an essential feature of a circulatory system.

≫ *Pointer*

The four-chambered heart consists largely of cardiac muscle, a specialised tissue that is capable of rhythmical contraction and relaxation of its own accord throughout a person's life. The heart muscle is said to be '**myogenic**'.

The heart–cardiac cycle

The heart

The heart consists of a relatively thin-walled collection chamber and a thick-walled pumping chamber which are partitioned into two, allowing the complete separation of oxygenated and deoxygenated blood. The heart is, in effect, two pumps side by side.

The cardiac cycle

The cardiac cycle describes the sequence of events in one heartbeat. The pumping action of the heart consists of alternating contractions (**systole**) and relaxations (**diastole**).

There are three stages to the cardiac cycle:

Stage 1 – The right and left ventricles relax, the tricuspid and bicuspid valves open as the atria contract and blood flows into the ventricles.

Stage 2 – The atria relax and the right and left ventricles contract together forcing blood out of the heart into the pulmonary artery and the aorta as the semilunar valves are opened. The tricuspid and bicuspid valves are closed by the rise in ventricular pressure. The pulmonary artery carries deoxygenated blood to the lungs and the aorta carries oxygenated blood to the various parts of the body.

Stage 3 – The ventricles relax and the pressure in the ventricles falls. Blood under high pressure in the arteries causes the semilunar valves to shut, preventing blood from going back into the ventricles. Blood from the vena cavae and pulmonary veins enters the atria and the cycle starts again.

The following describes the flow of blood through the left side of the heart.

The left atrium is relaxed and receives oxygenated blood from the pulmonary vein. When full, the pressure forces open the bicuspid valve between the atrium and ventricle. Relaxation of the left ventricle draws blood from the left atrium. The left atrium contracts, pushing the remaining blood into the right ventricle through the valve. With the left atrium relaxed and with the bicuspid valve closed the left ventricle contracts. The strong muscular walls exert a strong pressure and push blood away from the heart through the semilunar valves through the pulmonary arteries and the aorta.

- Both sides of the heart work together, i.e. both ventricles contract at the same time, both atria contract together. One complete contraction and relaxation is called a heartbeat.

- After contraction, and the compartment has been emptied of blood, it relaxes, to be filled with blood once more.

- The ventricles contain more muscle than the atria and so generate more pressure to force the blood a greater distance.

- The left ventricle has a thicker muscular wall than the right ventricle as it has to pump the blood all round the body, whereas the right ventricle has only to pump the blood a shorter distance to the lungs.

Pressure changes in the circulatory system

- The highest pressure occurs in the aorta/arteries that show a rhythmic rise and fall corresponding to ventricular contraction.
- Friction with vessel walls causes a progressive drop in pressure. Arterioles have a large total surface area and a relatively narrow bore causing a substantial reduction from aortic pressure. Their pressure depends on whether they are dilated or contracted.
- The extensive capillary beds have a large cross-sectional area. These beds create an even greater resistance to blood flow.
- There is a relationship between pressure and speed and the pressure drops further due to leakage from capillaries into tissues.
- The return flow to the heart is non-rhythmic and the pressure in the veins is low but can be increased by the massaging effect of muscles.

Control of heartbeat

- In the wall of the right atrium is a region of specialised cardiac fibres called the sinoatrial node (SAN) which acts as a pacemaker.
- A wave of electrical stimulation arises at this point and then spreads over the two atria causing them to contract more or less at the same time.
- The electrical stimulation is prevented from spreading to the ventricles by a thin layer of connective tissue. This acts as a layer of insulation (it is important that the muscles of the ventricles do not start to contract until the muscles of the atria have finished contracting).
- The stimulation reaches another specialised region of cardiac fibres, the atrio-ventricular node (AVN), which lies between the two atria and which passes on the excitation to specialised tissues in the ventricles.
- From the AVN the excitation passes down the **Bundle of His** to the apex. The Bundle branches into **Purkinje tissue**, fibres in the ventricle walls which carry the wave of excitation upwards through the ventricle muscle.
- The impulses cause the cardiac muscle in each ventricle to contract simultaneously from the apex upwards.

Control of heartbeat.

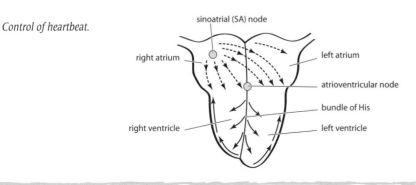

Key Terms

The Bundle of His = a strand of modified cardiac muscle fibre continuous with the AVN and Purkinje tissue fans over the walls of the ventricles.

Purkinje tissue = a network of fibres in the wall of the ventricles.

Grade boost

The graphical analysis of pressure changes in the heart is a favourite exam question. Be prepared to describe the pressure changes involved in the flow of blood from one chamber of the heart to another, together with the associated opening and closing of the valves.

Grade boost

Don't confuse the cardiac cycle with the control of heartbeat.

 quickfire

㉗ What is the significance of the heart pumping from the base upwards?

Blood

Blood is a tissue made up of cells (45%) in a fluid plasma (55%).

- Plasma is made up largely of 90% water, with soluble food molecules, waste products, hormones, plasma proteins, mineral ions and vitamins dissolved in it.
- Blood cells which are of two main types:
  - Erythrocytes or red blood corpuscles.
  - Leucocytes or white blood corpuscles.

quickfire

㉘ Give features of the red blood cells that make them well suited for the carriage of oxygen.

Functions of blood

- Plasma – transports carbon dioxide, digested food products, hormones, plasma proteins, fibrinogen, antibodies, etc., and also distributes heat.
- Red blood cells – are filled with the pigment haemoglobin, are biconcave in shape and do not contain a nucleus. Their function is the carriage of oxygen.
- White blood cells – these are of two groups:
  - Granulocytes, which are phagocytic, have granular cytoplasm, lobed nuclei and engulf bacteria.
  - Agranulocytes, which produce antibodies and antitoxins, have clear cytoplasm and spherical nuclei.

The transport of oxygen

Red blood cells load (pick up) oxygen in the lungs where the partial pressure is high and the haemoglobin becomes saturated with oxygen. The cells carry the oxygen as oxyhaemoglobin to the respiring tissues, e.g. muscle, where the partial pressure is low (as oxygen is being used up in respiration to create energy). Oxyhaemoglobin then unloads its oxygen, that is, it dissociates.

A saturation graph of haemoglobin with increasing partial pressure of oxygen may be expected to show a straight line bisecting the two axes but, in fact samples of haemoglobin exposed to increasing partial pressures of oxygen show an oxygen dissociation curve.

quickfire

㉙ List two ways in which the shape of the haemoglobin dissociation curve line differs from the theoretical line.

Oxygen dissociation curve.

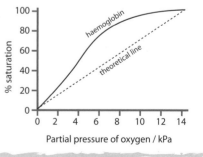

An efficient respiratory pigment readily picks up oxygen at the respiratory surface and releases it on arrival at the tissues. Respiratory pigments have a high affinity for oxygen when the concentration is high but this is reduced when the concentration is low. The special properties of haemoglobin ensure that at high **partial pressures of oxygen**, as in the lungs, it combines readily with large amounts of the gas, i.e. haemoglobin becomes almost fully saturated with oxygen.

At higher partial pressure of carbon dioxide the oxygen dissociation curve shifts to the right. This phenomenon is known as the **Bohr Effect**. When oxygen reaches respiring tissues, such as muscle, the high partial pressure of carbon dioxide there enables haemoglobin to unload its oxygen even more readily.

To summarise:

- When the respiratory pigment haemoglobin is exposed to a gradual increase in oxygen tension, it absorbs oxygen rapidly at first but more slowly as the tension continues to rise. This relationship is known as the oxygen dissociation curve.
- The more the dissociation curve of haemoglobin is displaced to the right, the less readily it picks up oxygen, but the more easily it releases it.
- The more the dissociation curve of haemoglobin is displaced to the left, the more readily it picks up oxygen, but the less readily it releases it.
- The release of oxygen from haemoglobin is facilitated by the presence of carbon dioxide, when the partial pressure of oxygen is high, as in the lung capillaries, oxygen combines with the haemoglobin to form oxyhaemoglobin.
- When the partial pressure of oxygen is low, as found in the respiring tissues, then the oxygen dissociates from the haemoglobin.
- When the partial pressure of carbon dioxide is high, haemoglobin is less efficient at taking up oxygen and more efficient at releasing it.

The dissociation curve of foetal haemoglobin

The blood of the foetus and the mother flow closely together in the placenta but rarely mix. To enable the foetal haemoglobin to absorb oxygen from the maternal haemoglobin in the placenta the foetus has haemoglobin that differs (in two of the four polypeptide chains) from the haemoglobin of the adult. This structural difference makes the foetal haemoglobin dissociation curve shift to the left of that of the adult. The foetal haemoglobin combines with oxygen more readily than does the mother's haemoglobin. That is, the foetal haemoglobin has a greater affinity for oxygen.

Key Terms

Bohr Effect = at higher partial pressures of carbon dioxide the oxygen curve shifts to the right.

The partial pressure of oxygen (pO_2) = a measure of the oxygen concentration. The greater the concentration of dissolved oxygen, the higher its partial pressure.

›› Pointer

The quantity of oxygen carried by haemoglobin depends not only on the partial pressure of oxygen but also the partial pressure of carbon dioxide.

›› Pointer

The Bohr Effect is concerned with the uptake of oxygen by haemoglobin at low partial pressures and the offloading of oxygen at the tissues as a result of the high levels of carbon dioxide there.

‹‹‹‹ quickfire

30 Explain the significance of the difference between the oxygen dissociation curve of a foetus and the mother.

Grade boost

The oxygen dissociation curve is a difficult concept so study it thoroughly.

quickfire

③ Suggest one change which could be observed in the blood of an athlete who had been training at high altitude.

Transport of oxygen in other animals

The chemical composition of haemoglobins is not the same in all animals. Some animals have become adapted to living in habitats where there are low levels of oxygen.

- The lugworm has a low metabolic rate and lives in the sand on the seashore (worm casts can be seen at low tide). The lugworm pumps seawater through its burrow, giving access to the limited amount of dissolved oxygen present. To enable it to load the oxygen more readily it has haemoglobin with a dissociation curve very much to the left compared with a human haemoglobin dissociation curve.

- With increase in altitude there is a drop in atmospheric pressure. This is significant for animals, such as the llama, because the partial pressure of oxygen in the atmosphere is less at high altitude. To compensate for this:
  - The llama possesses haemoglobin which loads more readily with oxygen in the lungs. Haemoglobin of this sort has a dissociation curve to the left of normal haemoglobin.
  - At high altitude the number of red cells in the blood of mammals increases.

Pointer

Myoglobin acts as an energy store in muscle.

Myoglobin

Myoglobin is far more stable than haemoglobin and its dissociation curve is far to the left of haemoglobin. At each partial pressure of oxygen, myoglobin has a higher percentage oxygen saturation than haemoglobin. Normally, the respiring muscle obtains its oxygen from haemoglobin. However, if the oxygen partial pressure becomes very low, as when exercising, the oxymyoglobin unloads its oxygen.

Chloride shift – the transport of carbon dioxide

Carbon dioxide is transported in blood cells and plasma in three ways:

- 5% in solution in the plasma (this is inadequate to meet the needs of most organisms).
- 85% as hydrogencarbonate.
- 10% in combination with haemoglobin to form carbamino-haemoglobin.

The following describes a series of reactions known as the chloride shift:

- Carbon dioxide diffuses into the red blood cell (RBC) and combines with water to form carbonic acid. The reaction is catalysed by carbonic anhydrase.
- Carbonic acid dissociates into H^+ and HCO_3^- ions. HCO_3^- ions diffuse out of the RBC into the plasma where they combine with Na^+ ions from the dissociation of sodium chloride to form sodium hydrogen carbonate.
- H^+ ions provide the conditions for the oxyhaemoglobin to dissociate into oxygen and haemoglobin.
- H^+ ions are buffered by their combination with haemoglobin and the formation of haemoglobinic acid (HHb).
- The oxygen diffuses out of the RBC into the tissues.
- To balance the outward movement of negatively charged ions, chloride ions diffuse in.
- This is known as the chloride shift and it is by this means that the electrochemical neutrality of the RBC is maintained.

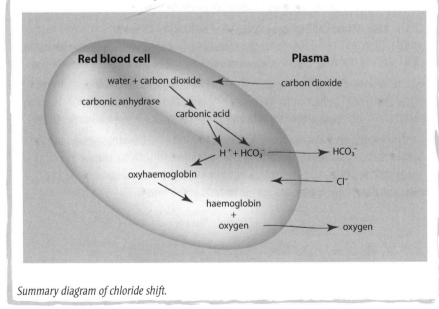

Summary diagram of chloride shift.

>> *Pointer*

Some carbon dioxide is transported in the red blood cells but most is converted in the red blood cells to bicarbonate, which is then dissolved in the plasma. The chloride shift refers to the influx of chloride ions into the red blood cells to preserve electrical neutrality.

 Grade boost

Haemoglobin acts as a buffer helping to maintain the blood pH by removing hydrogen ions from solution.

 quickfire

③② What is the significance of the inward movement of chloride ions into the red blood cell?

STILL TO REVISE!

Intercellular fluid

quickfire

㉝ Name the two opposing forces involved in forcing tissue fluid out of the blood plasma in capillaries and into the surrounding tissues.

quickfire

㉞ Name the two routes by which tissue fluid returns to the bloodstream.

The capillaries are the site of exchange between the blood and the cells of the body. They are well adapted to allow the exchange of materials between the blood and the cells:

- They have thin, permeable walls.

- They provide a large surface area for exchange of materials.

- Blood flows very slowly through the capillaries allowing time for exchange of materials.

- Blood consists of the fluid plasma that carries the blood cells, dissolved materials and large molecules, called plasma proteins. The blood is contained in a closed system but fluid from the plasma escapes through the walls of the capillaries. This fluid is called **tissue fluid** and bathes the cells, supplying them with glucose, amino acids, fatty acids, salts and oxygen. The tissue fluid also removes waste materials from the cells.

- The factors responsible for the movements of solutes and water into and out of the capillaries are blood pressure and diffusion.

- When blood reaches the arterial end of a capillary it is under pressure because of the pumping action of the heart and the resistance to blood flow of the capillaries. This hydrostatic pressure forces the fluid part of the blood through the capillary walls into the spaces between the cells.

- This outward flow is opposed by the reduced water potential of the blood, created by the presence of the plasma proteins.

- The hydrostatic pressure of the blood is greater than the osmotic forces, so there is a net flow of water and solutes out of the blood.

- At the arterial end of the capillary bed the diffusion gradient for solutes, such as glucose, oxygen and ions, favours movement from the capillaries to the tissue fluid. This is because these substances are being used during cell metabolism.

- At the venous end of the capillary bed the blood pressure is lower and water passes into the capillaries by osmosis. The reduced water potential of the blood created by the presence of the plasma proteins causes a net inflow of water.

- At the venous end, tissue fluid picks up CO_2 and other excretory substances. Some of this fluid passes back into the capillaries, but some drains into the lymphatic system and is returned eventually to the venous system via the thoracic duct, which empties into a vein near the heart.

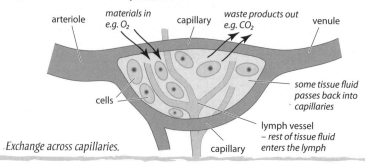

Exchange across capillaries.

Transport in plants

The xylem and phloem are distributed differently in roots and stems.

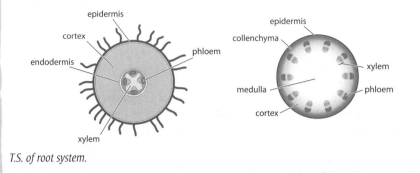

T.S. of root system.

The structure of xylem

Xylem is made up of four different types of cells: vessels, tracheids, fibres and xylem parenchyma.

Vessels and tracheids are dead cells and they form a system of tubes through which water can travel. The cells are dead because lignin has been deposited on the cellulose cell walls rendering them impermeable to water and solutes. These cells also provide mechanical strength and support to the plant.

Water uptake by the roots

The large quantities of water lost through the stomata of the leaves by the process of transpiration must be replaced with water from the soil. The region of greatest uptake is the root hair zone where the surface area of the root is enormously increased by the presence of root hairs.

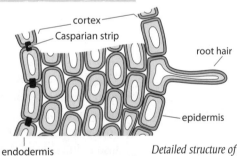

Detailed structure of part of root.

The soil water contains a very weak solution of mineral salts and so has a high water potential. The vacuole of the root hair cell contains a strong solution of dissolved substances and has a low water potential. Water passes into the root hair cell down a water potential (WP) gradient from a high WP to a low WP by osmosis. Water can travel across the cells of the cortex of the root along three pathways:

- The apoplast – through the cell wall.
- The symplast – through the cytoplasm and plasmodesmata.
- The vacuolar pathway – from vacuole to vacuole.

However, it is considered that the two main pathways are the symplast and apoplast pathways. Most of the water probably follows the latter pathway as this is the faster of the two.

Grade boost

It is important to know whether processes involving water transport are active or passive.

③⑦ Explain why water passing along the apoplast pathway has to be re-routed when it reaches the endodermis.

⟫ Pointer

A water potential gradient exists across the cortex. The WP is high in the root hair cell and lower in the adjacent cells.

The xylem tissue is found in the centre of the root and is surrounded by a single layer of cells called the endodermis. The cell walls of the endodermis are impregnated with a waxy material called

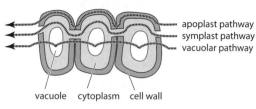

Three routes for water transport across cortex.

suberin. This forms a distinctive band known as the Casparian strip. The suberin is waterproof and the Casparian strip prevents the use of the apoplast pathway. The only way that water can pass across the endodermis to the xylem is along the symplast pathway. There is some evidence that salts may then be actively secreted into the vascular tissue from the endodermal cells. This makes the WP in the xylem more negative, causing water to be drawn in from the endodermis, so promoting the movement of water into the xylem from the cortex. The water potential gradient produced creates a force known as the root pressure.

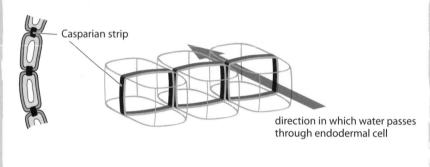

Endodermal cell showing the Casparian strip.

Uptake of minerals

Generally, minerals are taken up by the root hairs by active transport from the soil solution. Once absorbed, the mineral ions may move along the apoplast pathway carried in solution by the water being pulled up the plant in the transpiration stream. When minerals reach the endodermis, the Casparian strip prevents further movement along the cell walls. The ions enter the cytoplasm of the cell, from where they diffuse or are actively transported into the xylem. For example, nitrogen usually enters the plant as nitrate ions/ammonium ions which diffuse along the concentration gradient into the apoplast stream but enter the symplast by active transport against the concentration gradient and then flow via plasmodesmata in the cytoplasmic stream. At the endodermis, ions must be actively taken up to by-pass the Casparian band, which allows the plant to selectively take up the ions at this point.

③⑧ Suggest the effect of a respiratory inhibitor on mineral uptake.

The movement of water from roots to the leaves

- Water travels in the xylem up through the stem to the leaves, where most of it evaporates from the internal leaf surface and passes out, as water vapour, into the atmosphere.
- The transpiration of water from the leaves draws water across the leaf from the xylem tissue along the same three pathways as in the root.
- As water molecules leave xylem cells in the leaf, they pull up other water molecules. This pulling effect is known as the transpiration pull and is possible because of the large **cohesive** forces between water molecules and the **adhesive** forces which exist between the water molecules and the hydrophilic lining of the vessels. These two forces combine to maintain the column of water in the xylem.
- The theory of the mechanism by which water moves up the xylem is known as the Cohesion-Tension theory.
- **Capillarity** is another force that may contribute to the rise of water in the xylem. Water rises up narrow tubes by capillary action but this force is probably of more relevance in small plants than large trees.

Key Terms

Cohesion = water molecules tend to stick together.

Adhesion = the water molecules stick to the walls of the xylem.

Capillarity = the tendency for water to rise in narrow tubes.

›› Pointer

There are three main forces involved in water transport from root to leaf: root pressure, which is a 'push', transpiration, a 'pull' and capillarity.

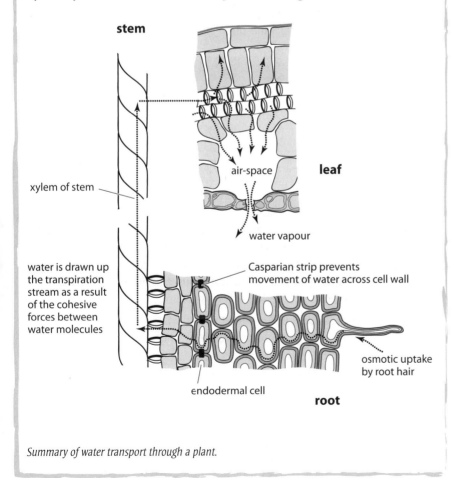

stem

xylem of stem

air-space   **leaf**

water vapour

water is drawn up the transpiration stream as a result of the cohesive forces between water molecules

Casparian strip prevents movement of water across cell wall

osmotic uptake by root hair

endodermal cell

root

Summary of water transport through a plant.

Grade boost

It is incorrect to state that the cuticle prevents water loss, it reduces water loss.

≫ *Pointer*

The rate of transpiration is measured using a potometer. It actually measures the rate of water absorption but if the cells of the plant are fully turgid, the rate of water absorption and the rate of transpiration are the same.

Grade boost

Using the idea of 'diffusion shells' can help to explain the factors affecting the rate of transpiration. In still air the shells remain on the leaf surface but wind will blow them away increasing the water potential gradient.

Transpiration

A major problem confronting all terrestrial organisms is how to avoid desiccation. Land plants, in particular, are continually losing water vapour to the atmosphere. In fact, about 99% of the water absorbed by the plant can be lost by evaporation through the leaves. This evaporation of water from inside the leaves through the stomata to the atmosphere is known as transpiration and gives rise to the transpiration stream. All plants have to balance water uptake with water loss. If a plant loses more water than it absorbs, it wilts. If a plant loses an excessive quantity of water, it reaches a point where it cannot regain its turgor and it dies.

Plants face a dilemma. The stomata need to be open during the day to allow the exchange of gases between the tissues of the leaf and the atmosphere. However, the presence of pores in the leaves means that valuable water is lost from the plant. Most of the water is lost through the stomata, although around 5% of the total water vapour loss can occur through the leaf epidermis. This loss is normally reduced due to the presence of the waxy cuticle on the surface of the leaves.

The rate at which water is lost from plants is called the transpiration rate and is dependent on external factors such as temperature, humidity and air movement. Any factor that increases the water potential gradient between the water vapour in the leaf and the surrounding atmosphere increases the rate of transpiration.

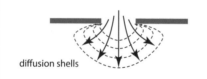
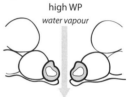

high WP
water vapour

diffusion shells

low WP in atmospheric air

Movement of water out of leaf.

- Temperature – a rise in temperature provides additional kinetic energy for the movement of water molecules. This additional energy accelerates the rate of evaporation of water from the walls of the mesophyll cells and, if the stomata are open, speeds up the rate of diffusion of water vapour into the surrounding atmosphere. The water potential of the atmosphere becomes lower as its temperature is raised and it can hold more moisture.

- Humidity – the air inside a leaf is saturated with water vapour but the humidity of the atmosphere surrounding a leaf varies, with values exceeding 70% being rare in Britain. Thus the water potential gradient between leaf and atmosphere is always great and when the stomata are open, water vapour rapidly diffuses from the leaf.

- Air movement – transpiration in still air results in the accumulation of a layer of saturated air at the surface of leaves. This offers considerable resistance to the diffusion of water vapour through stomata and thus reduces the rate of transpiration. Movement of the surrounding air reduces the thickness of the layer of saturated air and results in increased transpiration.

- Light intensity also affects transpiration by controlling the degree of stomatal opening.

In reality, these factors do not act independently but interact with each other, e.g. more water is lost on a dry, windy day than on a humid, still day. This is because the sub-stomatal chamber has a high water potential as the walls of the spongy mesophyll cells are saturated with water. The water evaporates from the walls and moves down a gradient of water potential from the plant to the atmosphere, which has a low percentage relative humidity; the wind having reduced the thickness of the layer of saturated air at the leaf surface.

Mesophytes, xerophytes and hydrophytes

Plants can be classified on the basis of structure, in relation to the prevailing water supply, into three groups:

- Hydrophytes (water plants).
- Xerophytes (plants that live in conditions where water is scarce).
- Mesophytes (plants living in conditions of adequate water supplies).

Most land plants growing in temperate regions belong to the latter category. Usually the water that they lose by transpiration is readily replaced by uptake from the soil, so they do not require any special means of conserving water. If such a plant loses too much water, the plant wilts and the leaves droop. The leaf surface area is reduced and photosynthesis becomes less efficient.

Mesophytes

Mesophytes flourish in habitats with adequate water supply. Most plants of temperate regions are mesophytes and, most importantly, most of our crop plants are mesophytes. They are adapted to grow best in well-drained soils and moderately dry air. Mesophytes lose a lot of water but excessive loss is prevented by closure of the stomata. Water uptake during the night makes good the water lost during the day.

Mesophytes need to survive unfavourable times of the year, particularly when the ground is frozen:

- Many trees and shrubs shed their leaves before the onset of winter.
- The aerial parts of many non-woody plants die off as a result of frost or cold winds but their underground organs survive, e.g. bulbs, corms.
- Most annual mesophytes (plants that flower, produce seed and die in the same year) survive the winter as dormant seeds.

Grade boost

Study graphs showing how environmental factors affect the rate of transpiration.

quickfire

(39) For each of the following conditions state whether the rate of transpiration increases or decreases:
1. Wind speed increases
2. Light intensity increases
3. Humidity increases.

Grade boost

quickfire

40 Explain how sunken stomata reduce transpiration.

quickfire

41 State one modification to reduce water loss shared by plants and insects.

Xerophytes

Xerophytes are plants which show xeromorphic adaptations. These plants have adapted to living in conditions of low water availability and have developed modified structures to prevent excessive water loss. They may live in hot, dry desert regions, cold regions where the soil water is frozen for much of the year, exposed, windy locations.

Ammophila arenaria (marram grass) is an example of a xerophyte which colonises sand dunes around the coast. The sand dune habitat makes it difficult for a mesophyte to survive there because there is no soil, rapid drainage of rain water occurs, there are high wind speeds, salt spray and a lack of shade from the sun.

Marram grass shows the following modifications:

- Rolled leaves – large thin-walled epidermal cells at the bases of the grooves shrink when they lose water from excessive transpiration, causing the leaf to roll onto itself. This has the effect of reducing the leaf area from which transpiration can occur.
- Sunken stomata – stomata are found in grooves on the inner side of the leaf. They are located in pits or depressions so that water vapour is trapped outside the stomata. This reduces the water potential gradient between the leaf and the atmosphere and so reduces the rate of diffusion of water.
- Hairs – stiff, interlocking hairs trap water vapour and reduce the water potential gradient.
- Thick cuticle – the cuticle is a waxy covering over the leaf surface which reduces water loss. The thicker this cuticle the lower the rate of cuticular transpiration.

Hydrophytes

Hydrophytes grow submerged or partially submerged in water. An example is the water lily, which is rooted to the mud at the bottom of a pond and has floating leaves on the surface of the water. Hydrophytes are adapted as follows:

- As water is a supportive medium, they have little or no lignified support tissue.
- Surrounded by water there is little need for transport tissue, so xylem is poorly developed.
- Leaves have little or no cuticle.
- Stomata are found on the upper surface of the leaves.
- Stems and leaves have large air spaces, forming a reservoir of oxygen and carbon dioxide. These gases also provide buoyancy to the plant tissues when submerged.

Translocation

Structure of phloem

Phloem is a living tissue and consists of several types of cells, the main ones being sieve tubes and companion cells. The sieve tubes are the only components of phloem obviously adapted for the longitudinal flow of material. They are formed from cells called sieve elements placed end to end. The end walls do not break down but are perforated by pores. These areas are known as sieve plates. Cytoplasmic filaments containing phloem protein extend from one sieve cell to the next through the pores in the sieve plate. The sieve tubes do not possess a nucleus and during their development most of the other cell organelles disintegrate. Each sieve tube element is closely associated with at least one companion cell, which has dense cytoplasm, large centrally placed nuclei, many mitochondria, and they are connected to the sieve tube element by plasmodesmata.

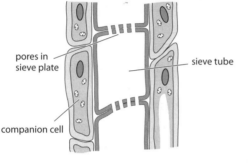

pores in sieve plate

sieve tube

companion cell

L.S. phloem tissue.

Transport in the phloem

Experimental evidence suggests that the phloem is the tissue concerned with the translocation of organic substances. Early evidence was obtained from ringing experiments where cylinders of outer bark tissue (removing the phloem) were removed from woody stems and the contents of the phloem above and below the cylinder were later analysed. More recently, the technique of radioactive tracing with labelled metabolites using aphids has been used. An aphid has a hollow, needle-like mouthpart (stylet), which is inserted into sieve tubes to feed on the sap. To sample the phloem sap, the aphid is anaesthetised and the stylet is cut off, leaving it attached to the plant. As the sap is under pressure it exudes from the very fine tube and can be collected and analysed. These experiments have enabled scientists to demonstrate that translocation is a rapid process, much too rapid to be explained by diffusion.

>> Pointer

The products of photosynthesis are transported in the phloem, away from the site of synthesis in the leaves (the 'source'), to all the other parts of the plant where they are used for growth or storage (the 'sink'). In plants the transport of the soluble organic materials, sucrose and amino acids, is known as translocation.

quickfire

㊷ Name the two main types of cells in the phloem.

quickfire

㊸ Suggest why there are many mitochondria in companion cells.

» Pointer
The transport of water is a passive process, whereas the transport of organic solutes is an active process.

Radioisotope labelling is a technique where carbon dioxide labelled with radioactive carbon is supplied to an illuminated plant leaf. The radioactive carbon is fixed in the sugar produced in photosynthesis and its translocation to other parts of the plant can be traced using autoradiography. The 'source' leaf and 'sink' tissues are placed firmly on photographic film in the dark for 24 hours. When the film is developed, the presence of radioactivity in parts of the tissue shows up as 'fogging' of the negatives. The technique shows that the sugar is transported in both upward and downward directions, since the radioactivity is shown in the aerial parts of the plant as well as the roots. An autoradiograph of a transverse section of a stem of a treated plant shows fogging only where phloem was in contact with the film.

Theories of translocation

Grade boost
You are not required to give details of any of the theories.

The main theory put forward to explain the transport of organic solutes is known as the mass flow hypothesis (1937). This theory suggests that there is a passive mass flow of sugars from the phloem of the leaf where there is the highest concentration (the source) to other areas, such as growing tissues, where there is a lower concentration (the sink).

Arguments against the theory include:

- It does not explain the existence of the sieve plates which seemingly act as a series of barriers impeding flow.
- Sucrose and amino acids have been observed to move at different rates and in different directions in the same tissue.
- Phloem tissue has a relatively high rate of oxygen consumption, and translocation is slowed down or stopped altogether if respiratory poisons such as potassium cyanide enter the phloem.
- The companion cells contain numerous mitochondria and produce energy but the mass flow hypothesis does not suggest a role for the companion cells.

Recent theories put forward suggest:

- An active process may be involved.
- The observation of streaming in the cytoplasm of individual sieve tubes could be responsible for bi-directional movements along individual sieve tubes, providing there was some mechanism to transport solutes across the sieve plates.
- Some scientists have observed protein filaments passing through the sieve pores and suggest that different solutes are transported along different filaments.

At present scientists do not have a complete understanding of the mechanism of translocation and a great deal of debate continues.

Reproductive strategies

Reproduction is the ability to produce other individuals of the same species and is a fundamental characteristic of living things. New individuals may be produced by either asexual or sexual reproduction, or in some species by both methods. In animals asexual reproduction is far less common than it is in plants, protoctists and prokaryotes.

>> *Pointer*
Bacteria are also able to reproduce sexually but you would not be required to provide details.

Asexual and sexual reproduction

In plants and animals reproduction is achieved in two ways. Some organisms, such as bacteria and yeast reproduce asexually; more advanced animals reproduce only sexually; flowering plants can reproduce both asexually and sexually.

- *Asexual reproduction* – this method rapidly produces large numbers of individuals having an identical genetic composition. A group of genetically identical offspring produced by this method is called a clone. Examples of asexual methods in animals include binary fission and budding. Plant examples are: bulbs, e.g. daffodil; runners, e.g. strawberry; tubers, e.g. potato.
- *Sexual reproduction* – this method usually involves two parents, is less rapid than asexual and produces offspring that are genetically different. Diploid body cells produce haploid sex cells or gametes. The fusion of haploid gametes is always involved.

44 Many organisms can reproduce sexually and asexually. Give one advantage and one disadvantage of sexual reproduction.

The advantages and disadvantages of the two methods of reproduction

In asexual reproduction lack of variety is a disadvantage in adapting to environmental change but its main advantage is that if an individual has a genetic makeup suited to a particular set of conditions, large numbers of this successful type may be built up.

In sexual reproduction:

- There is an increase in genetic variety so enabling a species to adapt to environmental change.
- It allows the development of a resistant stage in the life cycle, which enables the species to withstand adverse conditions.
- The formation of spores, seeds, and larvae enables the dispersal of offspring. This reduces intra-specific competition and so enables genetic variety to develop as required.

Mutations, although rare, help to create a little variety in asexual reproduction. Mutations arise more frequently (but still rarely) during sexual reproduction because of the greater complexity of the process.

Grade boost

Throughout this topic consider the transition made by vertebrates from an aquatic environment to a terrestrial existence.

quickfire

45 State two advantages of internal fertilisation.

Gamete production and fertilisation

Gamete production

- Living organisms have **diploid** body cells and **haploid** sex cells or gametes.
- Body cells with the full chromosome number are produced by mitosis.
- Haploid cells with half the chromosome number are produced by meiosis.
- At fertilisation the haploid sperm fuses with a haploid egg to produce a diploid fertilised egg. The **zygote** formed then divides many times by mitosis to grow into a new individual.
- Males and females usually produce different-sized gametes. The male gamete is small and extremely motile and the female gamete is large and sedentary, normally due to the presence of stored food.
- Mammalian eggs differ in that they contain very little stored food and instead the materials for development are obtained from the maternal blood supply through the placenta.

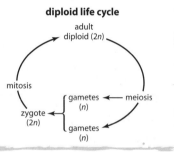

diploid life cycle

Fertilisation

Many aquatic organisms discharge their gametes directly into the sea or freshwater. As the gametes are quickly dispersed by the water, there is a strong possibility that many eggs will not encounter sperm. In animals this sort of fertilisation is known as external fertilisation. There is a considerable wastage and so large numbers of gametes of both sexes have to be produced. In the frog, the joining of sperm and egg is facilitated by a sexual coupling. When the eggs are laid by the female, the male immediately releases seminal fluid over them. While amphibians, in general, can move efficiently on land, many must return to water to reproduce and fertilisation is external.

In most terrestrial animals, however, fertilisation occurs inside the body of the female and this is called internal fertilisation. Generally, this requires the use of some kind of intromittent organ to introduce the sperm into the female's body.

Internal fertilisation has several advantages. There is less chance of gametes being wasted. It allows the male gamete to become independent of the need for water for movement. The fertilised egg can be enclosed within a protective covering before it leaves the female's body. This is what happens in animals that lay eggs. Some animals take this idea further and the embryos develop within the female parent and derive nourishment from her. This reaches its greatest development in those mammals that nourish their developing young before birth by means of a placenta.

Adaptations of animals to life on land

In many animals the fertilised egg or zygote undergoes development outside the body of the parent. The developing offspring are easy prey for predators and provide food for other species. Many eggs are produced to ensure that at least a few survive. In insects, although fertilisation is internal, the fertilised eggs are normally laid on a suitable food source and the embryo develops outside the body. Internal fertilisation ensures that all the sperm are deposited in the female's reproductive tract.

The gradual adaptation to life on land includes the evolution of an amniote egg in reptiles and birds. The egg has a fluid-filled cavity surrounded by a membrane outside which is a protective shell that encloses the embryo within the yolk sac. Birds incubate eggs and the embryo completes its development outside the mother's body.

In mammals the young are retained for a considerable time in the mother's womb or uterus but there is no shell. The embryo is nourished there from the mother's blood supply via the placenta. The young are born in a relatively advanced state of development.

Parental care

Many animal species reproduce by laying fertilised eggs that are left to develop unattended. There is little or no parental care. Pronounced parental care is typical of most species of birds and mammals. It includes the provision of shelter from unfavourable environmental conditions, feeding, protection from predators, and in some species the training of their offspring as they prepare for adult life. Generally, the more parental care provided, the fewer the number of offspring produced.

Insects are a group that have successfully colonised the land

During the development of the zygote, an intermediate form, referred to either as the nymph or the larva, is formed. These are juvenile forms that develop from the egg stage in the life cycle. Insects have a hard outer exoskeleton and in order to grow they have to shed the skin. They do this several times during their development. They are said to undergo an incomplete metamorphosis where the young nymph, which resembles the adult, hatches from the fertilised egg and goes through a series of moults until it reaches full size.

More advanced insect species develop from a larval stage that is quite different from the adult. The process involves considerable changes and is called a complete metamorphosis. These insects, such as the butterfly and housefly, also have an additional stage called the pupa or chrysalis. The larva hatches from the egg and is specialised for feeding and growing. The larva undergoes a period of change within the pupa and emerges as the adult that is specialised for dispersal and reproduction.

quicKfire

46 State two adaptations of animals to life on land.

Incomplete metamorphosis

Complete metamorphosis

» Pointer

It is a significant advantage to have a pupal stage in the life cycle of an insect for two reasons; it allows the insect to overcome unfavourable conditions as well as enabling the development of a more specialised adult.

Grade boost

You need only have a superficial knowledge of plant reproduction at this level.

≫ Pointer

The evolution of flowering plants was linked to their association with insects for pollination.

quickfire

47 State three reasons why flowering plants have been so successful in the colonisation of the land.

Colonisation of land by Angiosperms

Simple plant forms such as algae, e.g. seaweeds, are confined to an aquatic environment for at least part, if not all, of their lives. Other plant groups, such as mosses and ferns, are confined to damp areas, as the male gametes require a surface film of water in which to swim to the egg. Like the successful land animals, the conifers and flowering plants became independent of water for their reproduction and so were able to colonise the land.

Flowering plants are well suited to life on land because of their method of reproduction and because they have efficient water-carrying xylem vessels which also function in support. Flowers have pollen grains with a hard coat to withstand desiccation. These contain the male gamete which can be transferred to the female part of the plant.

Pollen grains can be transferred by wind or insects. Plants such as grasses have small, green inconspicuous flowers and the pollen is carried by wind. In plants with brightly coloured flowers and scent for attraction, the pollen is carried by insects. The male gamete travels through the tissue of the female part to the egg by means of a pollen tube. This means that sexual reproduction no longer depends on gametes having to travel through a film of water to reach the egg cell. Following fertilisation, the fertilised egg develops into a seed containing a food store.

The flowering plants are the most successful of all terrestrial plants. There are more than 300,000 species. They are found in every type of habitat. A key feature of their success is their relationship with animals, e.g. plants attract animals, particularly insects, to their flowers in order to feed and so exploit their mobility for pollination and seed dispersal. Another important development was the enclosure of the eggs in an ovary and the evolution of the seed. Seeds that result from fertilisation contain food reserves and have a resistant coat and so are able to withstand adverse conditions.

Why did the flowering plants become so successful?

1. The interval between flower production and the setting of seed is usually a matter of weeks.

2. The production of the seed with a food store enables the embryo to develop until leaves are produced above ground and are able to carry out photosynthesis. The seed also protects the embryo from desiccation and other hazards.

3. Generally, leaves are deciduous and succulent and decay rapidly on falling to the ground. This enables humus to be produced and as a consequence the rapid recycling of ions for reuse by plants.

Adaptations for nutrition

Nutrition is the process by which organisms obtain nutrients to provide energy to maintain life functions, and matter to create and maintain structure. Organisms have evolved several different methods of obtaining nutrients.

Methods of nutrition

Autotrophic nutrition

Living organisms that can make their own food are called autotrophs. They provide food for all other life forms and so they are also known as producers. There are two types of autotrophic nutrition:

- Photosynthesis is the process by which green plants build up complex organic molecules, such as sugars, from carbon dioxide and water. The source of energy for this process comes from sunlight, which is absorbed by chlorophyll and related pigments. Algae and certain types of bacteria can also photosynthesise using energy from sunlight.

- Chemosynthesis is a process carried out by autotrophic bacteria. They use the energy derived from special methods of respiration to synthesise organic food.

Heterotrophic nutrition

Heterotrophs cannot make their own organic food. They have to consume complex organic food material produced by autotrophs. Since they eat or consume ready-made food they are known as consumers. All animals are consumers and are dependent on producers for food. Heterotrophs include animals, fungi, some types of protoctists and bacteria.

There are a number of different types of heterotrophic nutrition:

- Holozoic feeders include nearly all animals. They take their food into their bodies and break it down by the process of digestion. Most carry out this process inside the body within a specialised digestive system. The digested material is then absorbed into the body tissues and used by the body cells. Animals that feed solely on plant material are termed herbivores, those that feed on other animals are carnivores, and detritivores are animals that feed on dead and decaying material.

- Saprophytes are also known as saprobionts, and include all fungi and some bacteria. They feed on dead or decaying matter and do not have a specialised digestive system. They feed by secreting enzymes such as proteases, amylases, lipases and cellulases onto the food material outside the body and then absorb the soluble products across the cell membrane by diffusion. This is known as extracellular digestion. Microscopic saprophytes are called decomposers and their activities are important in the decomposition of leaf litter and the recycling of valuable nutrients, such as nitrogen.

Grade boost

Learn the terms for the different methods of nutrition as you will come across them again later in this section.

48 Name two groups of microscopic saprobionts.

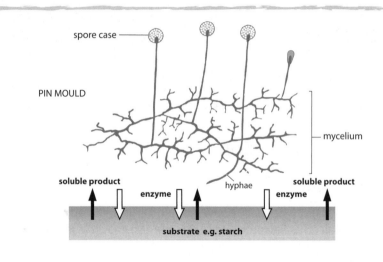

Saprophytic nutrition in a fungus.

- Parasites are organisms that feed on another living organism, referred to as the host. Some parasites live in the body of the host while others live on the surface. The host always suffers harm to some degree and often death. Parasites are considered to be very highly specialised organisms and show considerable adaptations to their particular way of life. Examples of parasites are the tapeworm, potato blight (caused by a fungus), and *Plasmodium* spp., the malarial parasite.

- Mutualism is also known as symbiosis and involves a close association between members of two different species, but in this case both derive some benefit from the relationship, e.g. the digestion of cellulose by micro-organisms in the gut of an herbivore. Cows and sheep feed mainly on grass, a high proportion of which is made up of cellulose cell walls. Like other herbivores, cows and sheep do not secrete the enzyme cellulase and so cannot digest cellulose. Instead they have mutualistic bacteria which live in a specific part of the specialised stomach (the rumen). These bacteria produce the enzymes for them and in return the bacteria gain other digestive products and suitable conditions for growth.

Processing food in the digestive system

Organic molecules must be broken down by digestion and absorbed into the body tissues from the digestive system before utilisation in the body cells. Digestion and absorption take place in the gut, which is a long, hollow, muscular tube. The gut is organised to allow movement of its contents in one direction only. In simple organisms which feed on only one type of food, the gut is undifferentiated. However, in more advanced organisms with a varied diet, the gut is divided into various parts along its length and each part is specialised to carry out particular steps in the processes of mechanical and chemical digestion as well as absorption.

In the human, the main regions of the gut are: the mouth, oesophagus (gullet), stomach, small intestine (duodenum, ileum), large intestine and anus. The food is processed as it passes along the various regions of the gut. It is propelled along the gut by the process of **peristalsis.**

- Ingestion is the taking of food into the body through the mouth.

- Digestion is the breakdown of large, insoluble food molecules into simple, soluble molecules by means of enzymes. Mechanical digestion in humans is achieved by the cutting and/or crushing action of the teeth followed by the rhythmical contractions of the gut. The gut wall, particularly the stomach, has layers of muscle to fulfil this function. These are responsible for mixing the food and pushing it along the gut. The physical action also has an important role as it increases the surface area over which enzymes can act. The chemical action of digestion is achieved through the secretion of digestive enzymes.

- Absorption is the passage of digested food through the gut wall into the blood.

- Egestion is the elimination from the body of food that cannot be digested, e.g. cellulose cell walls of plants.

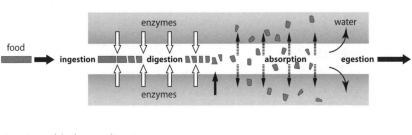

Functions of the human digestive system.

Key Term
Peristalsis = waves of muscular contraction.

 **quickfire**

(49) Define digestion.

» Pointer

You should always give functions to various parts of structures. The outer layer serves to protect, the muscle layer propels food along the gut, the sub-mucosa contains blood vessels to supply nutrients and nerves to co-ordinate activities, the inner layer produces secretions.

quickfire

(50) Name the four layers of the gut wall.

The structure of the human gut

Throughout its length from the mouth to the anus the gut wall consists of four tissue layers surrounding a cavity (lumen) of the gut.

- The outer serosa consists of a layer of tough connective tissue that protects the wall of the gut and reduces friction from other organs in the abdomen as the gut moves during the digestive process.

- The muscle layer consists of two layers of muscle running in different directions, the inner circular muscle and the outer longitudinal muscle.

- Collectively these muscles cause waves of muscular contractions, peristalsis, which propels food along the gut. Behind the ball of food the circular muscles contract and the longitudinal muscles relax, thus helping move the food along.

- The sub-mucosa consists of connective tissue containing blood and lymph vessels to take away absorbed food products as well as nerves that co-ordinate the muscular contractions involved in the process of peristalsis.

- The mucosa is the innermost layer and lines the wall of the gut. It secretes mucus which lubricates and protects the mucosa. In some regions of the gut this layer secretes digestive juices, in others it absorbs digested food.

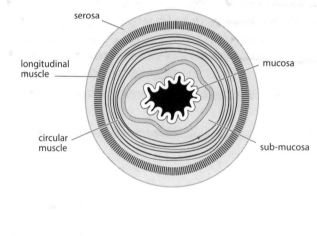

General structure of the gut wall.

Digestion

The absorption of nutrients by the gut epithelial cells is only possible if the large molecules, carbohydrates, fats and proteins are first broken down or digested into smaller products by means of enzymes. Different enzymes are required to carry out the digestion of the different food substrates and usually more than one type of enzyme is needed for the complete digestion of a particular food.

- Carbohydrates (polysaccharides) are first broken down into disaccharides and then into monosaccharides. The enzyme amylase hydrolyses starch to the disaccharide maltose but another enzyme, maltase, is required to break down the maltose to the monosaccharide, glucose.
- Proteins are broken down into polypeptides, then dipeptides, and finally into amino acids.
- The general name given to the protein-digesting enzymes is peptidase.
- Proteins are extremely large molecules so endopeptidases hydrolyse peptide bonds within the protein molecule and exopeptidases hydolyse peptide bonds at the ends of these shorter polypeptides.
- Fats are broken down to fatty acids and glycerol by just one enzyme, lipase.

Regional specialisations of the mammalian gut

- *The mouth* – mechanical digestion begins in the mouth when food is chewed using the teeth. The food is also mixed with saliva from the salivary glands. Saliva is a watery secretion, containing mucus and salivary amylase, together with some mineral ions which help to keep the pH in the mouth slightly alkaline, the optimum pH for amylase. Saliva is important for lubricating the food before it is swallowed. Amylase breaks down starch to maltose. After chewing, the ball of food is swallowed and mucus lubricates its passage down the oesophagus.
- *The stomach* – food enters the stomach and is kept there by the contraction of two rings of muscles, one at the stomach entrance and one at the junction with the duodenum. Food may stay in the stomach for up to four hours and during this time the muscles of the stomach wall contract rhythmically and mix up the food with gastric juice secreted by glands in the stomach wall. Gastric juice contains acid that gives the stomach contents a pH of 2.0. As well as providing the optimum pH for the enzymes, the acid kills most bacteria in the food. Gastric juice also contains peptidase enzymes which hydrolyse the protein to polypeptides. Mucus is important in forming a lining to protect the stomach wall from the enzymes and acid as well as assisting in the movement of food within the stomach.

Grade boost

Throughout this section highlight the digestion of carbohydrate, fats and proteins in three different colours.

quickfire

⑮ State the function of
1. Bile
2. Mucus.

⟫ Pointer

Explain how the different regions of the gut achieve different pH values. For example, the walls of the duodenum secrete an alkaline juice.

- *The small intestine* – the small intestine is divided into two regions, the duodenum and the ileum. Relaxation of the muscle at the base of the stomach allows small amounts of the partially digested food into the duodenum a little at a time. The duodenum makes up the first 20cm of the small intestine and receives secretions from both the liver and the pancreas.

- Bile is produced in the liver and stored in the gall bladder from where it passes into the duodenum via the bile duct. It contains no enzymes but the bile salts are important in emulsifying the lipids present in the food. Emulsification is achieved by lowering the surface tension of the lipids, causing large globules to break up into tiny droplets. This enables the action of the enzyme lipase to be more efficient as the lipid droplets now have a much larger surface area. Bile also helps to neutralise the acidity of the food as it comes from the stomach.

- The pancreatic juice is secreted from the exocrine glands in the pancreas and enters the duodenum through the pancreatic duct. It contains a number of different enzymes:
  - Endopeptidases, which hydrolyse protein to peptides.
  - Amylase, which breaks down any remaining starch to maltose.
  - Lipase, which hydrolyse lipids into fatty acids and glycerol.

The walls of the duodenum contain glands that secrete an alkaline juice and mucus. The alkaline juice helps to keep the contents of the small intestine at the correct pH for enzyme action, and the mucus is for lubrication and protection.

Enzymes secreted by cells at the tips of the villi complete digestion:

- Maltase hydrolyses maltose into two glucose molecules.
- Endopeptidases and exopeptidases complete the digestion of polypeptides to amino acids.

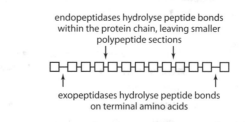

endopeptidases hydrolyse peptide bonds within the protein chain, leaving smaller polypeptide sections

exopeptidases hydrolyse peptide bonds on terminal amino acids

Endo- and exopeptidase

The end products of carbohydrate digestion are all monosaccharides. The final stage of carbohydrate digestion is intracellular, as disaccharides are absorbed by the plasma membrane of the epithelial cells before being broken down into monosaccharides.

Absorption

The ileum is well adapted for absorption:

In humans the ileum is very long and the lining is folded to give a large surface area compared to a smooth tube. On the folds are numerous finger-like projections called villi. On the surface of the villi are epithelial cells with microscopic projections called microvilli. These increase the surface area of the cell membrane of the epithelial cells for absorption.

Absorption follows digestion and takes place mainly in the small intestine. Because energy is required for active absorption, the epithelial cells also contain large numbers of mitochondria.

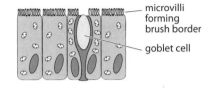

microvilli forming brush border

goblet cell

Structure of epithelial cells lining the small intestine.

- Glucose and amino acids are absorbed across the epithelium of the villi by a combination of diffusion and active transport. They pass into the capillary network that supplies each villus. The blood from the venules, which contains the dissolved food, eventually reaches the hepatic portal vein and and is carried to the liver.

- Fatty acids and glycerol are passed into the lacteal. This is a blindly ending lymph capillary found in the centre of each villus. Fatty acids and glycerol are transported in the lymphatic system, which ultimately opens into the blood stream at the thoracic duct.

To summarise, the following methods of transport occur:

- Fatty acids, glycerol and most vitamins pass through the membrane of the epithelial cells by diffusion.

- However, the glucose, amino acids and dipeptides require energy in the form of ATP for absorption by active uptake. Dipeptides are then digested intracellularly into simple amino acids. Glucose and amino acids then diffuse from the epithelial cell into the blood.

The large intestine

The large intestine is about 1.5 metres long and is divided into the caecum, the appendix, the colon and the rectum. Water and mineral salts are absorbed from the colon along with vitamins secreted by micro-organisms living in the colon. These bacteria are responsible for making vitamin K and folic acid. By the time it reaches the rectum, indigestible food is in a semi-solid condition. It consists of residues of undigested cellulose, bacteria and sloughed cells and passes along the colon to be egested as faeces. This process is called defecation.

>> **Pointer**

Glucose is absorbed from the blood by cells, for energy release in respiration.
Amino acids are absorbed for protein synthesis, excess cannot be stored so are deaminated whereby the removed amino groups are converted to urea and the remainder to carbohydrate and stored.
Lipids are used for membranes and hormones, excess are stored as fat.

quickfire

53 Into which parts of the villus are the end products of digestion absorbed?

Grade boost

The transport of digested products involves diffusion or active transport. Be sure which method is involved with specific products.

Adaptations to different diets

>> *Pointer*

Since food is retained for cutting, crushing, grinding or shearing according to diet, mammals have evolved different types of teeth with each type being specialised for a different function. Herbivores and carnivores have teeth specialised to suit their diets.

QUICKFIRE

54 State two differences that can be observed when comparing the dentition of the dog and the sheep.

Grade boost

In answering a question requiring a comparison, it is not sufficient to state, for example, that a dog has carnassials, you must also state that a sheep does not possess them.

Teeth

Humans have four different types of teeth: incisors, canines, pre-molars and molars. The teeth are not particularly specialised because humans are omnivores, that is, they eat they both plant and animal material. The teeth of herbivores and carnivores are specialised to perform particular functions.

Herbivore dentition

Plant food is a tough material and the teeth of herbivores are modified to ensure that food is thoroughly ground up before it is swallowed. A grazing herbivore, such as a cow or sheep, has incisors on the lower jaw only and cuts against a horny pad on the upper jaw. The canine teeth are indistinguishable from the incisors. A gap called the diastema separates the front teeth from the side teeth or premolars. The tongue operates in this gap moving the freshly-cut grass to the large grinding surfaces of the cheek teeth. The jaw moves in a circular grinding action in a horizontal plane. The cheek teeth interlock, like the letter W fitting into the letter M. With time the grinding surfaces become worn down, exposing the sharp-edged enamel ridges which further increase the efficiency of the grinding process. The teeth have open, unrestricted roots so that they can continue to grow throughout the life of the animal.

Carnivore dentition

Carnivorous mammals, such as tigers, have teeth adapted for catching and killing prey, cutting or crushing bones and for tearing meat. The sharp incisors grip and tear flesh from bone. The canine teeth are large, curved and pointed for seizing prey, for killing and also tearing flesh. The premolars and molars are for cutting and crushing. Carnivores have a pair of specialised cheek teeth, called carnassials, which slide past each other like the blades of gardening shears. The jaw muscles are well developed and powerful to enable the carnivore to grip the prey firmly and help in crushing bone. There is no side-to-side movement of the jaw, found only in herbivores, as this would lead to the jaw being dislocated when dealing with prey. The vertical jaw movement is greater than in herbivores allowing the jaw to open widely for capturing and killing prey.

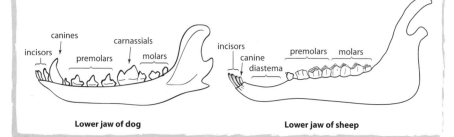

Lower jaw of dog   Lower jaw of sheep

Ruminants

Animals produce 35% of all the protein eaten by humans. Of this, nearly half comes from **ruminants**, such as cows and sheep, which eat mainly grass and forage, a large proportion of which consists of cellulose cell walls.

Mutualism or symbiosis involves a close association between members of two different species and where both organisms derive some benefit from the relationship. Herbivorous mammals such as cows and sheep lack the ability to produce cellulase enzymes and so cannot digest cellulose. A large proportion of plant material consists of cellulose cell walls. Certain herbivores, e.g. cows, have formed an association with cellulose-digesting bacteria which live in the gut of the cow. In this association the mammal acquires the products of cellulose digestion and the bacteria receive a constant supply of food and can grow in a suitable sheltered environment.

The cow provides a region of the gut for the bacteria to inhabit and in return the bacteria digest the cellulose for the cow. However, the region of the gut must be kept separate from the main digestive region so that:

- Food can be kept there long enough for the bacteria to carry out the digestion of the cellulose.
- The bacteria are isolated from the mammal's own digestive juices so that they are in the optimum pH for their activities and they are not killed by extremes of pH.

Ruminants have a 'stomach' which is made up of four chambers. Three of the chambers are derived from the lower part of the oesophagus and one chamber is the true stomach. Cellulose digestion takes place as follows:

- The grass is chopped by the teeth, mixed with saliva, and the cud formed is swallowed.
- In the rumen, the first region, the cud is mixed with cellulose-digesting bacteria to produce glucose. This is fermented to form organic acids that are absorbed into the blood, and provides energy for the cow. The waste products are carbon dioxide and methane that are passed out.
- The cud passes to the next region before being regurgitated into the mouth and chewed again.
- The cud passes directly into the third region where water is reabsorbed.
- The fourth and last region functions like a 'normal' stomach and protein is digested.
- The digested food passes to the next region, the small intestine, where the products of digestion are absorbed.

Structure of ruminant gut.

quickfire

55 Explain why it is necessary for the gut of a cow to contain cellulose-digesting bacteria.

>> **Pointer**

The gut of a carnivore is short, reflecting the ease with which protein is digested. However, the gut of a herbivore is long because digestion of plant material is difficult.

Grade boost

It is often useful to summarise a passage that you have read. For example, the gut region of a ruminant is adapted to its diet by having a long gut, containing cellulose-digesting bacteria and being able to regurgitate its food and 'chew the cud'.

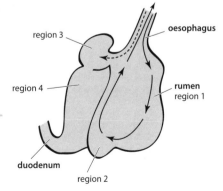

≫ Pointer

The tapeworm does not have a gut. This is because it is surrounded by the host's nutrients which are absorbed through the body surface.

⊙✗≪≪≪ quicKfire

56 Name two problems that the tapeworm has to overcome in the hostile environment of the gut of the host.

Parasitic nutrition

Parasites are organisms that live on or in another organism, called the host, and obtain nourishment at the expense of the host. Parasites therefore cause harm to some degree and often cause death.

Pork tapeworm – a parasite of the gut

All animals have a struggle to survive, to avoid competition with others and to avoid being preyed upon by other animals. Parasites have become specialised and undergone considerable evolutionary changes in order to survive in the host. The gut parasite (*Taenia solium*) is a particularly good example.

Imagine living in the gut of another animal! The tapeworm is ribbon-like and can be up to 10 metres long! It has a 'head' made up of muscle on which are suckers and hooks. Its body consists of a linear series of thin segments. The pork tapeworm has two hosts. The primary host is the human and the pig is the secondary host. The pig becomes infected if it feeds in drainage channels contaminated by human faeces. Humans are infected by eating undercooked infected pork.

Although the tapeworm lives in an immediate source of food, it needs to survive the hostile conditions found in the gut. The following are the problems that the gut parasite has to overcome in order to survive:

- It lives surrounded by digestive juices and mucus.
- Food, mixed with digestive juices, is in constant motion as it is churned about as well as being propelled along the length of the gut by peristaltic contractions of the muscular wall.
- It lives in extremes conditions of pH along the length of the gut.
- The immune system of the host.
- If the host dies then so does the parasite.

In order to survive the tapeworm must:

- Have a means of penetrating the host.
- Have a means of attachment to the host.
- Protect itself against the immune responses of the host.
- Develop only those organs that are essential for survival.
- Produce many eggs.
- Have an intermediate host.
- Have resistant stages to overcome the period away from a host.

The tapeworm has evolved the following structural modifications to enable it to live as a parasite:

- Suckers and a double row of curved hooks for attachment to the wall of the gut.

- A body covering which protects them from the host's immune responses.

- A thick cuticle and the production of inhibitory substances on the surface of the segments to prevent their digestion by the host's enzymes.

- Because they live in a stable environment they do not need to move around and do not require a sensory system. This has led to the degeneration of unnecessary organs. They do have a simple excretory and nervous system but most of the body is concerned with reproduction.

- The tapeworm is very thin and has a large surface area to volume ratio. It is surrounded by digested food so it has a very simple digestive system and pre-digested food can be absorbed over the entire body surface.

- Because the gut could not accommodate two tapeworms, each segment contains both male and female reproductive organs. Vast numbers of eggs are produced, with each mature segment containing up to 40,000 eggs. The mature segments pass out of the host's body with the faeces.

- The eggs have resistant shells and can survive until eaten by the secondary host. Further development can then take place and the embryos which hatch from the eggs move into the muscles of the pig and remain dormant until the meat of the pig is eaten by a human.

Harmful effects of the pork tapeworm

The adult worms cause little discomfort but, if the eggs are eaten by humans, dormant embryos form cysts in various organs and damage the surrounding tissue. Adults can be treated with appropriate drugs. Public health measures and frequent inspection of meat are essential measures.

◉ ⫸⫷⫷⫷⫷ quicKFire

㊼ Name two adaptations of tapeworms which are adaptations to their parasitic life.

≫ Pointer

The tapeworm is hermaphrodite, that is, both sex organs are present in the one individual. The gut could not accommodate two tapeworms so mating would be impossible. It therefore fertilises its own eggs.

Summary: Biodiversity and Physiology of Body System

Gaseous exchange

- Unicellular organisms use diffusion.
- In plants gases diffuse through stomata and to mesophyll cells of leaf.
- Multicellular organisms have:
  - a specialised exchange surface such as gills, lungs or trachea
  - ventilation mechanism
  - an internal transport system.

Evolution

- Biodiversity
- Natural selection
- Adaptive radiation
- Taxonomy
- Five kingdom classification
- Binomial system

Transport

Transport of materials in animals

- Transport of materials in animals involves:
  - vessels – arteries, veins and capillaries
  - a muscular pump – the heart
  - three main phases of heart beat – atrial systole, ventricular systole and diastole
  - heart muscle is myogenic and controlled by specialised regions of cardiac fibres.
- Blood is made up of plasma, red blood cells and white blood cells.
- Carbon dioxide is mainly carried in the form of hydrogen carbonate ions, and chloride shift maintains the electrochemical neutrality of the red blood cell.
- Red blood cells contain haemoglobin.
- Oxygen dissociation curve occurs when haemoglobin is exposed to a gradual increase in oxygen tension.
- White blood cells involved in body defence.

Transport of materials in plants

- Transport of water and mineral salts from the roots to the leaves.
- Transpiration of water from the leaves through the stomata.
- Forces involved include root pressure, cohesive and adhesive forces and transpiration pull.
- Xerophytes and hydrophytes have developed adaptations to grow best in their particular habitats.
- Transport of organic materials takes place in the phloem sieve tubes.
- Transport occurs from sources to sinks according to mass flow theory.

Nutrition

Types of nutrition

- Plants are autotrophic and can manufacture food by photosynthesis.
- Animals are heterotrophic and consume complex organic food material.
- Heterotrophic methods of feeding include saprobionts, parasites and mutualism.

In humans

- Food is processed by ingestion, digestion, absorption and egestion.
- Chemical digestion involves enzymes which hydrolyse large insoluble materials.
- The end products of digestion are glucose, amino acids, fatty acids and glycerol.
- The gut is divided into specialised regions for digestion and absorption.
- Digested products are absorbed in the ileum by diffusion, facilitated diffusion or by active transport.

Mammals have evolved adaptations

- Teeth of herbivores and carnivores are specialised to suit their diets.
- Ruminants have a specialised stomach in which bacteria live and digest cellulose.

The pork tapeworm is a parasite which has adapted to survive in the hostile conditions of the human gut.

Reproduction

- Advantages and disadvantages of asexual and sexual reproduction.
- Organisms have adapted to a terrestrial existence.
- Fertilisation may be:
  - external in aquatic organisms.
  - internal in terrestrial organisms.
- The embryo is protected from desiccation by:
  - an egg surrounded by a shell in reptiles and birds.
  - retention within the mother's body in mammals.
  - flowering plants and insects are particularly successful terrestrial groups.

Exam Practice and Technique

Exam practice and skills

WJEC AS Biology aims to encourage students to:

- develop their interest and enthusiasm for the subject, including developing an interest in further study and careers in the subject
- appreciate how society makes decisions about scientific issues and how the sciences contribute to the success of the economy and society
- develop and demonstrate a deeper appreciation of the skills, knowledge and understanding of How Science Works
- develop essential knowledge and understanding of different areas of the subject and how they relate to each other.

Examination questions are written to reflect the assessment objectives as laid out in the specification. Candidates must meet the following assessment objectives in the context of the content detailed in the specification.

Assessment objective AO1:
Knowledge and understanding of science and How Science Works

Candidates should be able to:

- recognise, recall and show understanding of scientific knowledge
- select, organise and communicate relevant information in a variety of forms.

47% of the questions set on the exam paper are recall of knowledge.

Assessment objective AO2:
Application of knowledge and understanding of science and How Science Works

Candidates should be able to:

- analyse and evaluate scientific knowledge and processes
- apply scientific knowledge and processes to unfamiliar situations including those relating to issues
- assess the validity, reliability and credibility of scientific information.

47% of the questions set on the exam paper include application of knowledge.

Assessment objective AO3:
How Science Works

Candidates should be able to:

- demonstrate and describe ethical, safe and skilful practical techniques and processes, selecting appropriate qualitative and quantitative methods
- make, record and communicate reliable and valid observations and measurements with appropriate precision and accuracy
- analyse, interpret, explain and evaluate the methodology, results and impact of their own and others' experimental and investigative activities in a variety of ways.

6% of the questions set on the exam paper include How Science Works.

BY1 and BY2: Written paper (1 hour 30 minutes)

The following is an approximate guide to the structure of the examination papers, BY1 and BY2:

Type of question	Marks per question	Number of questions per paper	
		BY1	BY2
Short structured	2–5	2–3	2–3
Longer structured	7–15	2–4	2–4
Essay (1 out of 2)	10	1	1
Total marks		70	70

Examination questions are written by the Principal examiner well in advance of the examination. A committee of experienced examiners discuss the quality of every question as well as the suitability of the wording.

The following is advice given in the specification:

The questions are worded very carefully so that they are clear, concise and unambiguous. Despite this, candidates tend to penalise themselves unnecessarily when they misread questions, either because they read them too quickly or too superficially. It is essential that candidates appreciate the precise meaning of each word in the question if they are to be successful in producing concise, relevant and unambiguous responses. The mark value at the end of each part of each question provides a useful guide as to the amount of information required in the answer.

Exam tips

Read the question carefully. Examiners try to make the wording of the questions as clear as possible but in the examination situation, it is all too easy to misinterpret a question. Read every word in every sentence carefully and use a highlighter pen if it helps you to focus on key words.

Understand the information

Only a certain percentage of the questions at AS are based on recall of knowledge. You may encounter unfamiliar material. It is important that you do not panic but think carefully and take your time and apply the principles that you have learnt to answer these types of question. You many also encounter a graph or table. Again, read the information carefully several times before you attempt the question.

Look at the mark allocation

Each question or part of a question is allocated a number of marks. You must make sure that if the question is worth three marks, then you must give three points to gain those marks.

Understand the instructions

Know the meaning of action words. Make sure that you are familiar with the terms below and that you understand what the examiner expects you to do.

Describe

This term may be used in a variety of questions where you need to give a step-by-step account of what is taking place. In a graph question, for example, you may be required to recognise a simple trend or pattern, then you should also use the data supplied to support your answer. At this level it is insufficient to state that the graph goes up and then flattens. You are expected to describe what goes up and by how much.

Explain

A question may ask you to describe and also explain. You will not be given a mark for merely describing what happens – a biological explanation is also needed.

Suggest

This action word often occurs at the end of a question when you are expected to put forward a sensible idea based on your biological knowledge. There may not be a definite answer to this question.

Name

This means that you must give no more than a one-word answer. You do not have to repeat the question or put your answer into a sentence. This is wasting time.

State

A brief, concise answer with no explanation.

Compare

If you are asked to make a comparison, do so. For example, if you are asked to compare the dentition of a cat and a sheep, don't write out two separate descriptions. Make a comparative statement, such as, 'a cat has carnassials a sheep does not'.

Tips about structured questions

Structured questions can be short, requiring a one-word response, or can include the opportunity for extended writing. The number of lined spaces on the exam question paper together with a mark allocation are indications of the length of answer expected and the number of points to be made. Structured questions are in several parts, usually about a common content. There is an increase in the degree of difficulty as you work your way through the question. The first part may be simple recall, perhaps defining a term, the most difficult part coming at the end of the question.

Tips about essays and diagrams

All too often candidates rush into essay questions, often writing everything they know about the topic without specifically answering the question. You should take your time reading the question carefully to discover exactly what the examiner

requires in the answer. Highlight the key words then write down a plan. This will not only help you to organise your thoughts but also give you a checklist to which you can refer back while writing your answer. In this way you will be less likely to repeat yourself, wander off the subject or miss out important points.

Consider this sample question:

> 'Describe the pathways and mechanism involved in the movement of carbon dioxide from the atmosphere to the palisade mesophyll cells of the leaf.'

The words to highlight are – describe, pathways, mechanism, carbon dioxide, atmosphere, palisade mesophyll cells.

The pathway refers to the route taken by the carbon dioxide through the stomata, air spaces of the spongy mesophyll layer and to the chloroplasts of the palisade mesophyll cells. Carbon dioxide moves in by diffusion because there is more carbon dioxide in the atmosphere than in the leaf during the day, since carbon dioxide is being used up in the process of photosynthesis.

When you have a plan to follow, it is so much easier to organise your thoughts while writing your answer.

Should you include a diagram in your essay answer?

Where appropriate you should include well-drawn, annotated diagrams. Even in essay questions this is an excellent way of communicating biology. In fact the rubric in the essay section states: 'Any diagrams in your answer must be fully annotated'. This means that you are encouraged to include a diagram but a labelled diagram is insufficient. 'Annotate' means adding a short description of the function or relevant point about the structure of the labelled part.

Questions and answers

This part of the guide looks at student answers to examination-style questions through the eyes of an examiner. There is a selection of questions on topics in the AS specification with two sample answers – one of a high grade standard and one of a lower grade standard in each case. The examiner commentary is designed to show you how marks are gained and lost so that you understand what is required in your answers.

BY1: Basic Biochemistry and Cell Structure

BY2: Biodiversity and Physiology of Body Systems

The diagrams show part of a molecule of starch (A) and part of a molecule of cellulose (B). The hexagonal shapes represent hexose sugars.

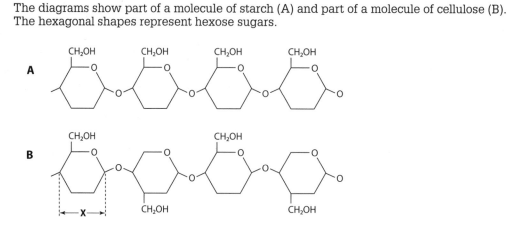

(a) Name monomer X and its form. *(1 mark)*
(b) Name the bond formed between two hexose sugars. *(1 mark)*
(c) State two structural differences between starch and cellulose. *(2 marks)*
(d) Starch is a compact storage polysaccharide. Cellulose has a structural role in plant cell wall. Describe how cellulose units are arranged in a complete molecule and how this arrangement gives cellulose a high tensile strength. *(2 marks)*

Tom's answer

(a) Glucose ⓧ ①

(b) Glycosidic ✔

(c) One has α glucose molecules and the other has β glucose. One is coiled and the other is straight chained. ⓧ②

(d) Cellulose is made up of long chains linked together like parallel strands in a rope. This makes it strong. ⓧ③

Examiner commentary

① Tom gains no mark as he has failed to give form, α, of glucose.

② Tom has failed to gain any marks as he has stated two correct differences but has not made a valid comparison.

③ Tom's statement is vague. He has also failed to use the information supplied in the diagram. He has the right idea but does not express himself in the detail required at this level.

Tom achieves 1 out of 6 marks.

Seren's answer

(a) α glucose ✔

(b) glycosidic ✔

(c) Starch is made up of α glucose molecules while cellulose consists of β glucose. ✔ Starch is branched and cellulose is unbranched. ✔ ①

(d) Starch is made up of coiled chains; cellulose has many long, parallel straight chains linked by hydrogen bonds. ✔ The adjacent molecules are rotated by 180°. ✔ A collection of parallel chains is called a microfibril. ②

Examiner commentary

① Seren has provided all the points required to gain the marks. Other possible answers include, starch is coiled and cellulose is straight chained; starch has 1–4 and 1–6 linkages, cellulose 1–4 linkages only; starch has two polysaccharides, cellulose one.

② To achieve a perfect answer Seren could have elaborated on the fact that rotating the adjacent molecules by 180° allows hydrogen bonds to be formed between –OH groups of parallel chains.

Seren achieves 6 out of 6 marks.

A   B

Q&A 2

The diagram (left) shows an organelle found in a liver cell.
(a) (i)   Name the organelle.   *(1 mark)*
    (ii) State the function of the organelle.   *(1 mark)*
    (iii) Name the structures labelled A and B in the diagram.
                                                                      (2 marks)
(b) Explain why the inner membrane is highly folded.
                                                                      (2 marks)
(c) Name the main molecule that is synthesised in this
    organelle.   *(1 mark)*
(d) Explain why liver cells have large numbers of these
    organelles present.   *(2 marks)*

Tom's answer

(a) (i) Mitochondrion ✓
    (ii) Respiration ✓
    (iii) A Highly folded inner membrane ✗
          B Cytoplasm ✗ ①
(b) To increase surface area ✓ ②
(c) ATP ✓
(d) Because ATP is needed to break down substances in the
liver, so mitochondria are required. ✗ ③

Examiner commentary

① Tom has failed to learn basic labelling of the organelle.

② Tom has gained one mark but has failed to note that the
question has been allocated a second mark – an increase
in surface area means that a maximum number of enzyme
molecules to be attached so more ATP can be produced.

③ Tom has not appreciated that the liver is a metabolically
active organ where many energy-requiring processes,
such as active transport, are taking place. Consequently, a
large amount of ATP is required.

Tom achieves 4 out of 9 marks

Seren's answer

(a) (i) Mitochondrion ✓
    (ii) It is the site of respiration ✓
    (iii) A crista ✓
          B matrix ✓
(b) To increase surface area for attachment of enzymes to
    produce ATP ✓ ✓
(c) ATP ✓
(d) Because the liver needs a large amount of ATP
    (energy) ✓ in order to carry out its function. The more
    mitochondria it has, the more ATP is synthesised. ①

Examiner commentary

① Seren knows that the liver needs a large amount of
ATP but has not stated that there are many energy-
requiring process taking place in the liver.

Seren achieves 8 out of 9 marks.

Q&A 3

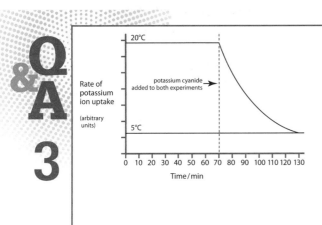

The graph shows the absorption of
potassium ions by young cereal plant root
hairs which were kept in aerated solutions
maintained at two different temperatures.
After 70 minutes potassium cyanide was
added to the solutions at each temperature.

(a) How does the information given show
    that the root hairs take up the ions by
    active transport?   *(3 marks)*
(b) Explain, why at low temperatures
    potassium uptake continues after the
    addition of potassium cyanide.   *(1 mark)*

Tom's answer

(a) The uptake is affected by temperature. ✓ The graph shows that adding cyanide makes the temperature go down from 20°C to 5°C. ⊗ ①

(b) Because active transport is still taking place. ⊗ ②

Examiner commentary

① Tom has not studied the graph carefully, particularly with regard to the axes. Nevertheless he gains one mark perhaps more by chance than good judgement! As the uptake is affected by temperature, he should immediately have thought of the involvement of enzymes. Active transport is an energy-requiring process. Cyanide inhibits enzymes involved in respiration and no ATP is produced; thus reducing the uptake of potassium ions.

② Limited uptake does occur by diffusion as it is a passive process unaffected by cyanide.

Tom achieves 1 out of 4 marks.

Seren's answer

(a) At 70 minutes the rate of uptake of potassium is reduced considerably until it reaches 130 minutes. Then it levels out and takes place slowly. Temperature causes the reduction ✓ because aerobic respiration is inhibited by the cyanide and there is no ATP for active uptake of potassium. ✓

(b) Diffusion still takes place and this is a much slower process than active transport. ✓ ①

Examiner commentary

① Although Seren has gained the mark, she could have stated that diffusion is a passive process and does not require energy. Uptake still occurs but at a much lower level as it is a slower process.

Seren achieves 4 out of 4 marks.

Q&A 4

The water potential of vacuolated cells is represented by the equation:

$$\Psi_{cell} = \Psi_s + \Psi_p$$

(a) Define the terms: *(2 marks)*
    (i) Osmosis   (ii) Water potential

(b) What is the water potential of pure water?   *(1 mark)*

(c) Two different cells P and Q are adjacent to one another in a plant. Calculate the missing values for each cell and complete the following table (all values are in kPa).   *(2 marks)*

Cell	Ψ_{cell}	Ψ_s	Ψ_p
P		-1200	+500
Q	-300		+300

(d) In which direction will water move between these two cells?   *(1 mark)*

Tom's answer

(a) (i) Osmosis is water moving from a high concentration to a low concentration across a semi-permeable membrane. ⊗ ①

   (ii) Water potential is solute potential and pressure potential. ⊗ ①

(b) Zero ✓

(c)

Cell	Ψ_{cell}	Ψ_s	Ψ_p
P	1700 ② ⊗	-1200	+500
Q	-300	0	+300

(d) Out of P ⊗ ③

Examiner commentary

① Tom has not learnt either definition and has missed out on easy opening marks. His definition of osmosis is not acceptable at this level. He has also attempted to use the equation to give an incorrect response.

② Tom lacks understanding of basic maths and has added together the two figures in the table.

③ As Tom has provided the incorrect answers in the table he is unable to give the correct direction of flow.

Tom achieves 1 out of 6 marks.

Seren's answer

(a) (i) This is the tendency for water to leave a system. ✓

(ii) This is the movement of water from a region where it has a high water potential to a region where it has a lower water potential through a partially permeable membrane. ✓

(b) 0 ✓

(c)

Cell	Ψ_{cell}	Ψ_S	Ψ_p
P	-700 ✓	-1200	+500
Q	-300	-600 ✓	+300

(d) From cell Q to cell P. ✓

Examiner commentary

Seren has a good understanding of the more difficult concept of osmosis and water potential. She has learnt her definitions and used the equation correctly to complete the table. She has also demonstrated that she understands that water moves from a high (or less negative) water potential to a low water (or more negative) potential.

Seren has achieved 6 out of 6 marks.

Q&A 5

Describe the methods of transportation across a cell membrane. *(10 marks)*

Mark scheme

A candidate would be expected to give any 10 of the following 15 marking points.

① *Diffusion* ② *definition* ③ *passive process/no energy or ATP involved* ④ *facilitated diffusion* ⑤ *requires use of a carrier/channel protein* ⑥ *osmosis* ⑦ *movement of water from high to low water potential* ⑧ *active transport* ⑨ *ATP required* ⑩ *against a concentration gradient* ⑪ *protein carriers in membrane required* ⑫ *endocytosis/exocytosis* ⑬ *energy-requiring process* ⑭ *pinocytosis of liquids/phagocytosis of solids* ⑮ *invagination/fusion with membrane*

Tom's answer

One method of transportation across a cell membrane is osmosis. ✓ ⑥ This is the net movement of water molecules from a region of high concentration to a region of low concentration across a semi-permeable membrane. ✗ ⑦ Another method is diffusion. ✓ ① This is the movement of molecules from a high concentration to a low concentration, e.g. oxygen diffuses from the high concentration in the air of the alveoli, across a membrane to the region of low concentration in the deoxygenated blood of the capillaries. ✓ ② Active transport is also a method of transportation ✓ ⑧ across a cell membrane and this method requires energy ✓ ⑨, e.g. in the root hair cells of a plant ions are actively transported into the cells from the soil solution.

Facilitated diffusion is a method of transport ✓ ④ that enables large substances, e.g. glucose, to pass across the cell membrane. These substances can fit into carrier proteins which then change shape taking the substance to the other side of the membrane. ✓ ⑪

Examiner commentary

⑦ Tom has not used the term 'water potential' in his answer and gains no mark.

Summative comment

Tom's answer lacks planning and organisation and his knowledge is very basic. He has also failed to mention all the answers required by the mark scheme.

Tom achieves 7 out of 10 marks.

Seren's answer

The methods that allow molecules to be transported across a cell membrane are diffusion, facilitated diffusion, osmosis, active transport, phagocytosis and pinocytosis.

Diffusion ✓ ① is the movement of molecules from where there are many to where there are few. ✓ ② The movement is random and will occur until equilibrium is reached. If the temperature is increased the particles have more energy, so they can therefore move faster and diffusion will occur at a higher rate. The concentration gradient is the difference in concentration between the two areas where particles are moving from and to. The distance that molecules need to travel also affects the rate because molecules move faster over smaller distances. Diffusion is a passive process. ✓ ③

Facilitated diffusion ✓ ④ is the same as diffusion, but in this case the particles being transported across the membrane travel through channel and carrier proteins. ✓ ⑤ Because the phospholipids that make up membranes have a layer of non-polar fatty acids inside them this means that only non-polar molecules can diffuse through. The larger polar molecules have to use an alternative route. Channel proteins allow these polar molecules through the membrane. Carrier proteins transport specific molecules and they can change shape to do so. Facilitated diffusion also follows the concentration gradient so it is therefore a passive process.

Osmosis ✓ ⑥ is similar to diffusion but it is specific to water. It depends on the water potential on either side of the membrane. Addition of solute molecules will lower the water potential, and areas with low water potential will draw in more water. ✗ ⑦ Osmosis is also a passive process.

Active transport ✓ ⑧ involves carrier proteins and because it takes place against a concentration gradient ✓ ⑩ it is an active process requiring energy in the form of ATP. ⑨

Phagocytosis is when a cell engulfs large molecules and brings them into a cell inside a membrane called a vesicle. Pinocytosis is the same thing but with smaller, soluble particles. ✓ ⑭ If a cell brings molecules in by this process it is called endocytosis ✓ ⑫ and if particles are being removed it is called exocytosis. Vesicles are produced at the Golgi body and released and move over to the cell membrane and fuse with it and secrete their contents out on the other side of the cell. ✓ ⑮

Examiner commentary

Seren provides a clear opening statement covering all the methods involved in transport across a cell membrane. She then methodically defines each method in turn and describes the relevant features of each. She also describes phagocytosis, pinocytosis and exocytosis. These methods are often overlooked by candidates answering this type of question.

⑦ Although Seren uses the term 'water potential', she has not clearly stated that osmosis is the movement of water molecules from a region of high water potential to a region of low water potential through a partially permeable membrane.

Summative comment

Seren has produced an excellent, well-planned essay. She begins by stating the methods involved in the transport of materials across a cell membrane. She then deals with each method in sufficient detail.

Seren achieves 10 out of 10 marks.

Q&A 6

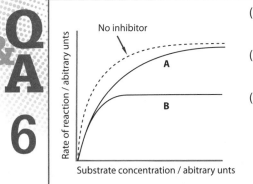

No inhibitor

A

B

Rate of reaction / abitrary unts

Substrate concentration / abitrary unts

(a) Apart from the presence of inhibitors and substrate concentration, state three factors that affect the rate of an enzyme-controlled reaction. *(3 marks)*

(b) Describe how the lock and key theory can be used to explain how an enzyme breaks down a substrate molecule. *(3 marks)*

(c) The graph shows how the rate of an enzyme-catalysed reaction varies with substrate concentration when affected by a competitive inhibitor and a non-competitive inhibitor.
  (i) Which line shows the competitive inhibitor? *(1 mark)*
  (ii) Give a reason for your choice. *(2 marks)*
  (iii) Explain how a competitive inhibitor works. *(3 marks)*

Tom's answer

(a) PH, ✓ temperature, ✓ amount of enzyme. ✗ ①

(b) The enzyme has an active site which has a very specific shape like the lock in a door. The shape is the same shape ✗ ② as the shape of the substrate, which fits just as a key fits into a lock, forming an enzyme–substrate complex. ✓ This then breaks down to form the products.

(c) (i) A ✓

  (ii) Non-competitive inhibition is not affected by the amount of substrate added. ✓ ③ With competitive inhibition, increasing the amount of substrate reduces the effect but the rate does reach its maximum eventually. ✓ ④

  (iii) The inhibitor has a similar shape to the active site and substrate ✗ ⑤ and so occupies the active site of the enzyme. This prevents the substrate from binding with the active site. ✓ So no reaction takes place.

Seren's answer

(a) pH, ✓ temperature, ✓ concentration of enzyme ✓

(b) The substrate fits into the active site of the enzyme ✓ ① because they have matching shapes. ✓ ① They form an enzyme-substrate complex and a product will be formed. ✓

(c) (i) A ✓

  (ii) Non-competitive inhibition is unaffected by the concentration of substrate added. ✓ However, with competitive inhibition increasing the concentration of substrate increases the likelihood of the substrate entering the active site with the formation ✓ ② of enzyme-substrate complexes.

  (iii) The inhibitor has a similar shape to that of the substrate and so occupies the active site of the enzyme in competition ✓ with the substrate. This prevents the entry of the substrate ✓ and prevents the formation of an enzyme-substrate complex. ✓

Examiner commentary

① Tom has stated the term 'amount' rather than 'concentration' of enzyme.

② The active site and the substrate do not have the same shape, they have matching shapes, i.e. the shapes are complementary.

③ Tom again uses the term, 'amount'. However the examiner will not penalise him again for the same error.

④ Tom has failed to state that an enzyme-substrate complex is formed. He has also failed to consider that the greater the substrate concentration, the greater the number of collisions between substrate and enzyme and therefore the greater the chance that the substrate will enter the active site.

⑤ Tom incorrectly states that the inhibitor has a similar shape to both substrate and active site and so fails to gain this mark.

Tom achieves 7 out of 12 marks.

Examiner commentary

① Although Seren has gained the mark here, she could have stated that the shape of active site and substrate are complementary to each other, rather than using the term 'matching'.

② Seren has grasped the principle that an increase in substrate concentration increases the chance of the formation of enzyme-substrate complexes. She could have added that this is due to an increase in the likelihood of a collision of the substrate with the active site rather than the inhibitor, as more substrate molecules are present.

Seren has an excellent grasp of this topic. She has a good understanding of how enzymes work and also how the rate of reaction is affected by inhibitors.

Seren achieves 12 out of 12 marks.

Q&A 7

(a) Describe and explain why the rate of reaction of an enzyme varies with temperature.
(b) What are immobilised enzymes? Describe the advantages of their use.   *(10 marks)*

Mark scheme

A candidate would be expected to give any 10 of the following 15 marking points.

① *Enzymes are inactive at 0°C* ② *With an increase in temperature the rate of reaction increases* ③ *an increase in temperature gives molecules greater kinetic energy* ④ *molecular collisions occur resulting in formation of enzyme-substrate complexes* ⑤ *the optimum temperature is 40°C.* ⑥ *Above 40°C enzymes are denatured* ⑦ *the increased vibration of molecules causes hydrogen bonds to break / tertiary structure is altered / shape of active site is altered. (Any 5 points out of 7)*

⑧ *Enzyme molecules that are fixed / bound / trapped* ⑨ *to an inert matrix / alginate bead* ⑩ *more stable at higher temperatures* ⑪ *can tolerate wider range of pH* ⑫ *easily recovered for reuse* ⑬ *several enzymes with different pH or temperature optima can be used at one time* ⑭ *reaction more easily controlled by adding or removing enzymes* ⑮ *specific so can select one type of molecule in a mixture / can be used for rapid detection of biologically important molecules. (Any 5 points out of 8)*

Tom's answer

As the temperature increases, so does the rate of reaction. ✓ ② Increasing the temperature means more energy ✗ ③ is available and this makes the enzyme and substrate molecules move about more. This means there are more collisions. ✗ ④ If the temperature is too high the enzyme is denatured. ✗ ⑥ An immobilised enzyme is an enzyme that is bound ✓ ⑧ in a inert matrix ✓ ⑨ e.g. gel capsules. There are many advantages of immobilised enzymes. They are cheap, they have a wide range over which they work. They have a wider range of temperature, pH ✓ ⑪, substrate concentration and enzyme concentration at which they work compared to a normal enzyme. They can also be reused easily. ✓ ⑫ They are used for testing for blood glucose level in biosensors. ✓ ⑮

Examiner commentary

③ ④ Tom refers to energy and substrate and enzymes moving more but fails to mention the word 'kinetic' and makes no reference to enzyme-substrate complexes.

⑥ He refers to 'denaturation' but does not qualify his statement with a reference to the temperature at which this begins to occur and offers no explanation as to what denaturation means.

Summative comment

Tom's answer to part (a) lacks detail. He fails to mention any specific temperatures and his answers are little above those expected at GCSE level. He gains maximum marks for part (b) despite poor expression.

Tom achieves 6 out of 10 marks.

Seren's answer

At 0°C enzymes are inactive ✓ ① but as the temperature rises, the rate of reaction increases. ✓ ② This is because the molecules gain kinetic energy and move about more. ✓ ③ This means that enzyme and substrate molecules are more likely to collide and these successful collisions result in the formation of enzyme-substrate complexes with the formation of a product. ✓ ④ 40°C is the optimum temperature for enzyme action. ✓ ⑤ Above this temperature enzymes start to denature. ✓ That is, enzyme molecules vibrate so much that the weak hydrogen bonds that hold the molecule together are broken and the tertiary structure is destroyed. ✓

Immobilised enzymes are enzymes trapped in an inert gel. ✓ ✓ ⑧⑨ They can be used over a larger range of temperatures and pH compared to normal enzymes. ✓ ⑪ This is because the inert gel helps maintain its tertiary structure. Because of the larger range of optima different immobilised enzymes can be used together. ✓ ⑬ They are good economically as the enzyme is easily separated from the product and can be reused. ✓ ⑫ Immobilised enzymes have a faster rate of reaction than ordinary enzymes and they are denatured at higher temperatures. The enzymes are used in biosensors for several different tests, one of which is blood glucose monitoring.

Examiner commentary

Seren describes specific temperatures which affect enzymes. She refers to the temperature at which enzymes are inactive and the optimum temperature of enzyme action. She does not simply define denaturation but also describes in molecular terms how it happens.

Seren has produced an excellent, concise and detailed answer.

Seren achieves 10 out of 10 marks.

Q&A 8

(a) Complete the table which compares DNA with messenger RNA (mRNA). *(4 marks)*

Feature	DNA	mRNA
Name of sugar		
Number of carbon atoms in sugar		
Number of polynucleotide chains in molecule		
Location in cell		

(b) The table below shows the relative amounts of the four bases in DNA taken from three sources.

Cellular source of DNA	Nitrogenous base (relative amounts)			
	Adenine	Guanine	Cytosine	Thymine
Rat muscle	28.6	21.4	21.5	28.4
Wheat seed	27.3	22.7	22.9	27.1
Yeast	31.3	18.7	17.1	32.9

(i) Explain why the relative amount of adenine is almost the same as the relative amount of thymine in each source. *(3 marks)*

(ii) Explain why the base sequence in a DNA sample taken from the bone marrow of a rat would be the same as that taken from the muscle of the same rat. *(3 marks)*

(iii) Explain how a sample of DNA from a rat sperm cell differs from that of a muscle cell from the same rat. *(3 marks)*

Tom's answer

(a)

Feature	DNA	mRNA
Name of sugar	Pentose	Pentose ⊗ ①
Number of carbon atoms in sugar	5	5 ✓
Number of polynucleotide chains in molecule	23	23 ⊗
Location in cell	nucleus	Cytoplasm ⊗

(b) (i) Because it all adds up to 100 so 50:50 A+G: C+T, e.g. 28.6 +21.4 = 50. ⊗ ②

(ii) The DNA throughout the whole rat will be the same, no matter where the sample is taken from. ⊗ ③

(iii) Because to make sperm (sex) cells it takes meiosis but meiosis ✓ makes only haploid non-identical cells, ✓ whereas mitosis would make muscle cells, for example, identical daughter cells. ⊗ ④

Examiner commentary

① Tom has made several errors in completing the table. He has failed to name the different sugars, deoxyribose and ribose present in the different nucleotides, and that DNA is double stranded whereas RNA is single stranded. He also fails to state that mRNA is present in both the nucleus and cytoplasm.

② Tom is asking the examiner to do his thinking for him! Complementary base pairing takes place and he has used abbreviations for the bases instead of writing their names in full. He fails to answer the question which is suggesting why the figures in the table for the base pairs are not the same. This is due to experimental error.

③ Tom has not mentioned mitosis and that replication of DNA takes place with the production of genetically identical body cells.

④ Tom correctly states that haploid sex cells are produced by meiosis. He also would have gained a mark had he gone back to (b)(ii) and included the information about mitosis.

Tom achieves 3 out of 13 marks.

Seren's answer

(a)

Feature	DNA	mRNA
Name of sugar	Deoxyribose	Ribose ✔
Number of carbon atoms in sugar	5	5 ✔
Number of polynucleotide chains in molecule	2	1 ✔
Location in cell	Nucleus	Nucleus and cytoplasm ✔

(b) (i) Because adenine joins with thymine in DNA by hydrogen bonds. This means that the same amount of adenine joins with the same amount of thymine. ✘ ✔ ①

(ii) The bone marrow has the same DNA as the muscle. This is because body cells are produced by mitosis ✔ so all cells are genetically identical. ✔ ②

(iii) Because the rat's sperm is haploid ✔ as sex cells are produced by meiosis. ✔ This produces variation in each sex cell, ✔ therefore one sperm may have different DNA to the other.

Examiner commentary

① Seren has failed to answer the question and has made a similar error to Tom. However, she has correctly named base pairs.

② Seren has failed to state that replication of DNA takes place during mitosis.

Seren achieves 10 out of 13 marks.

Q&A 9

(a) To which phylum do mammals belong? *(1 mark)*

(b) The drawings on the left show the bones in the limbs of three different mammals.

   (i) What name is given to limbs with the pattern of bones shown in the drawings? *(1 mark)*

   (ii) Suggest which of limbs A, B or C is best adapted for flight. *(1 mark)*

   (iii) Give the genus of the mammal with limb C. *(1 mark)*

(c) (i) The limbs of these mammals are similar in structure but serve quite different functions. What term is used to describe such structures? *(1 mark)*

   (ii) How are such structures used as evidence for evolution? *(1 mark)*

   (iii) State a biochemical technique used to confirm evolutionary relationships. *(1 mark)*

Tom's answer

(a) Vertebrates ✔ ①

(b) (i) Pentadactyl ✔   (ii) Limb B ✔   (iii) Phoca ✔

(c) (i) Adaptive radiation ✘ ②

   (ii) They have all got five digits ✘ ③

   (iii) Fingerprinting ✘ ④

Examiner commentary

① Tom has achieved a mark for his response but Chordata is a more correct term.

② Tom has used an incorrect term, the answer is 'homologous'.

③ He has made an incorrect guess.

④ At this level Tom must be more precise and include some form of DNA analysis.

Tom achieves 4 out of 7 marks.

Seren's answer

(a) Chordata ✔

(b) (i) Pentadactyl limb ✔

   (ii) Macroderma gigas ✔

   (iii) Phoca ✔

(c) (i) analogous ✘ ①

   (ii) It suggests that they have evolved from a common ancestor. ✔

   (iii) DNA hybridisation ✔

Examiner commentary

① Seren has confused the term analogous with homologous. Analogous refers to structures having the same function but a different origin. For example, the wings of a bird and an insect.

Seren achieves 6 out of 7 marks.

Q & A

10

The table gives some features of the cells of organisms from three different kingdoms.

Structure	Plant cell	Animal cell	Prokaryote cell
Mitochondrion			
Ribosome			
Cell wall			
Membrane-bound nucleus			

(a) (i) Complete the table by placing a tick in the appropriate box if the structure is present. Place a cross in the box if the structure is not present. *(4 marks)*

(ii) Give the names of the two kingdoms not listed in the table above. *(1 mark)*

(b) Write the following taxa in order of size starting with the largest group and ending with the smallest group. *(2 marks)*

Order, genus, phylum, class, family

Tom's answer

(a) (i)

Structure	Plant cell	Animal cell	Prokaryote cell
Mitochondrion	✓	✓	✗
Ribosome	✓	✓	✓
Cell wall	✓	✗	✓
Membrane–bound nucleus	✓	✓	✗

✓ ①

(ii) Fungi and Protozoa ✗ ②

(b) phylum order class family genus ✗ ✓ ③

Examiner commentary

① Although Tom has ticked the boxes correctly, he has failed to follow instructions and has not placed crosses in the other boxes. Consequently he gains only one mark.

② He has failed to name Protoctista and so gains no mark.

③ He gains one mark as he has correctly stated phylum as the largest taxon and genus as the smallest but he has the incorrect order between these two and fails to gain the second mark.

Tom achieves 2 out of 7 marks.

Seren's answer

(a) (i)

Structure	Plant cell	Animal cell	Prokaryote cell
Mitochondrion	✓	✓	✗
Ribosome	✓	✓	✓
Cell wall	✓	✗	✓
Membrane- bound nucleus	✓	✓	✗

✓ ✓ ✓ ✓

(ii) Fungi and Protoctista ✓

(b) phylum class order family genus ✓ ✓

Examiner commentary

This type of question would be considered an 'easy' opening question and Seren has made an excellent start.

Seren achieves 7 out of 7 marks.

Q & A

11

(a) Describe how organisms such as *Amoeba* and the earthworm are able to obtain oxygen without the need for specialised organs of gaseous exchange. *(5 marks)*

(b) Describe how pressure changes are achieved and bring about inspiration in a mammal. *(5 marks)*

Mark scheme

A candidate would be expected to give any 10 of the following 15 marking points.

① *Large surface area to volume ratio* ② *Amoeba has moist surface / earthworm epithelium or membrane / covered in mucus* ③ *earthworm has blood vessels close to surface of skin* ④ *reference to blood pigments* ⑤ *thin / permeable membrane* ⑥ *short diffusion path* ⑦ *low metabolic rate* ⑧ *external intercostal muscles contract* ⑨ *ribs move up and out* ⑩ *diaphragm contracts* ⑪ *volume of thorax increases* ⑫ *pressure in lungs decreases* ⑬ *atmospheric pressure is higher* ⑭ *air forced into lungs* ⑮ *reference to role of pleural membranes*

Tom's answer

(a) Amoeba is a unicellular organism and has a large surface area to volume ratio and can therefore satisfy the needs of the organism. ✓ ① It becomes flattened in shape to increase surface area allowing diffusion to take place over the whole of the body surface. ✗ ② This increases efficiency. There is very short distance between respiratory surfaces where diffusion takes place and the cells. ✗ ⑥ Amoeba is surrounded by water providing moist conditions for absorbing gases. This is why it does not require an internal transport system and blood pigments. Earthworm is a multicellular organism and has a larger volume than surface area and therefore requires specialised organs such as the lungs, as diffusion of gases is too slow a method to satisfy the organism's needs. However, the earthworm has a low metabolic rate ✓ ⑦ and has less demand for oxygen and also it becomes flattened in shape in order to increase the surface to volume ratio. Therefore it does not require specialised breathing organs such as lungs.

(b) The intercostal muscles flatten and the ribs move out. The pressure in the lungs gets less ✓ ⑫ and air rushes in.

Examiner commentary

② ⑥ Tom has confused the unicellular Amoeba with the multicellular flatworm. He then refers to the earthworm as being flat. He refers to Amoeba having a short diffusion pathway to the cells.

Summative comment

Tom is totally confused and has little understanding of the principles of gaseous exchange. His response to the question on inspiration is a reflection of the lack of effort made in his revision.

Tom achieves 3 out of 10 marks.

Seren's answer

(a) Amoeba is a single-celled organism. It has a high surface area to volume ratio. ✓ ① Diffusion is adequate enough to supply the cell with oxygen and remove carbon dioxide. Diffusion occurs over the whole surface of the cell. The cell has short diffusion paths to allow gases to diffuse easily. ✓ ⑦

The earthworm has a respiratory pigment, haemoglobin. ✓ ④ This has a high affinity for oxygen. The epidermis of the earthworm is covered in mucus and so is moist to allow easy diffusion of gases. ✓ ② The earthworm is a slow-moving animal, therefore has a low metabolism ✓ ⑦ and does not produce large amounts carbon dioxide, and does not require large amounts of oxygen. The earthworm has a capillary network which allows the circulation of blood. The capillary network is near to the epidermal surface of the worm ✓ ③ so is in close contact with the air to allow oxygen to diffuse in and be taken up by haemoglobin.

(b) Pressure changes are achieved because of the changes of volume in the thorax. The intercostal muscles contract ✓ ⑧ and the ribs move up and out, ✓ ⑨ the diaphragm contracts and flattens, ✓ ⑩ the volume of the thorax increases ✓ ⑪ and the pressure decreases in the thorax ✓ ⑫. The outside atmospheric pressure is greater than the pressure of the thorax so air is forced into the lungs. ⑭

Examiner commentary

Seren understands that diffusion is adequate for gaseous exchange in a unicellular organism but that the evolution of multicellular organisms necessitated the development of a circulatory system to transport oxygen from the body surface to the cells. She could have added that a blood pigment aids in the transport.

Summative comment

Seren achieves 10 out of 10 marks.

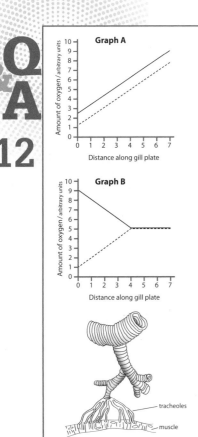

Graph A

Amount of oxygen / arbitrary units

Distance along gill plate

Graph B

Amount of oxygen / arbitrary units

Distance along gill plate

tracheoles

muscle

(a) Multicellular organisms like fish and mammals have part of their body surface modified to form specialised respiratory surfaces.

(i) List four properties that the respiratory surfaces of fish and mammals have in common. *(2 marks)*

(ii) Give two advantages to a mammal of possessing internal lungs. *(2 marks)*

(b) Both fish and mammals have ventilation mechanisms.

(i) Explain the function of ventilation mechanisms. *(2 marks)*

(ii) Name the muscles which operate the ventilation mechanisms in mammals. *(2 marks)*

(c) In some fish, water is forced over the gills in the opposite direction to the flow of blood in the gill plates (counter current flow). In others, water and blood flow in the same direction (parallel flow). The graphs show the relative amounts of oxygen in water and blood as the water moves across the gill plates for the different types of flow.

(i) Which line, dotted or solid, indicates the flow of blood? *(1 mark)*

(ii) Which graph shows parallel flow? *(1 mark)*

(iii) State which type of flow is the more efficient, giving a reason for your choice. *(2 marks)*

(d) Insects use a tracheal system for gas exchange as shown in the diagram.

(i) State the respiratory surface. *(1 mark)*

(ii) State two advantages of using a tracheal system for gas exchange. *(2 marks)*

Tom's answer

(a) (i) Large surface area, thin, permeable, blood supply. ✓ ✓

(ii) Protection ✓

(b) (i) To bring oxygen into the body ✗ ①

(ii) Diaphragm, intercostal muscles ✓ ✓

(c) (i) Solid line ✓   (ii) B ✓

(iii) Counter current flow because the whole surface is used. ✓ ✓

(d) (i) The tracheole ✗ ②

(ii) It has a large surface area ✗ ③

Examiner commentary

Tom has a fairly good understanding of this topic but some of his answers lack detailed reasoning.

① He has not given a full answer but gains one mark.

② Tom has failed to understand the principle that oxygen is sent directly to the tissues and does not appreciate that where the tracheole meets the cells is the respiratory surface.

③ Tom has guessed the answer and has failed to appreciate that the insect is able to direct oxygen to the tissues at a faster rate than a system involving blood.

Tom achieves 9 out of 15 marks.

Seren's answer

(a) (i) Large surface area, thin so a short diffusion pathway, permeable, good blood supply. ✓ ✓

(ii) Being inside the body means there is less water loss. The ribs also protect the lungs. ✓ ✓

(b) (i) A ventilation mechanism brings oxygen in and over the respiratory surface and then moves carbon dioxide out. This maintains a concentration gradient. ✓ ✓

(ii) Diaphragm, intercostal muscles ✓ ✓

(c) (i) Solid line ✓   (ii) B ✓

(iii) Counter current flow because this allows gas exchange to take place over all the gill plate and maintains a concentration gradient along the whole of the gill plate. ✓ ✓

(d) (i) Where the tracheole touches the muscle. ✓

(ii) Oxygen is supplied directly to the tissues and so no transport system or pigment is needed. This makes it a much faster system. ✓ ✓

Examiner commentary

Seren appreciates that in fish, having a concentration gradient between the water and blood along the whole length of the gill plate increases the efficiency of diffusion, and that in a different environment insects have evolved a unique method of transporting oxygen directly to the cells without the need for a transport system.

Seren achieves 15 out of 15 marks.

Q & A

13

Describe how the structure of leaves of flowering plants is adapted for gaseous exchange and photosynthesis.

(10 marks)

Mark scheme

A candidate would be expected to give any 10 of the following 15 marking points.

① *Leaves have a large surface area to absorb light.* ② *Leaves can orientate to expose maximum area to light.* ③ *Leaves are thin to allow light to penetrate to lower layers.* ④ *Cuticle is transparent to allow light to penetrate to mesophyll.* ⑤ *Palisade cells are elongated to reduce number of cross walls which would absorb light.* ⑥ *Palisade cells packed with chloroplasts to increase light absorption.* ⑦ *Chloroplasts can circulate (cyclosis) to absorb maximum light.* ⑧ *Spongy mesophyll cells are moist / large surface area for gas exchange.* ⑨ *Xylem transports water / phloem transports sugar produced.* ⑩ *Leaves are thin to reduce diffusion pathway.* ⑪ *Air spaces in spongy mesophyll for circulation of gases.* ⑫ *Stomatal pores permit entry and exit of gases / intercellular spaces allow gaseous exchange.* ⑬ *Waxy cuticle on upper surface reduces water loss.* ⑭ *Stomatal pores in lower epidermis reduce water loss.* ⑮ *Guard cells can close stomatal pores to reduce water loss.*

Tom's answer

The structure of leaves in flowering plants first begins with a top layer of epidermis cells. They have a waxy cuticle on top which is waterproof ✓ ⑭ and reduces the loss of water. These cells are also very transparent ✓ ④ which enables the light to pass through to the second layer of cells which is the palisade mesophyll. These cells are packed full with chloroplasts which are used in photosynthesis. ✗ ⑥ The third layer is the spongy mesophyll which is well suited for the passing of molecules to the palisade mesophyll. ✗ ⑧ There are many gaps between these cells which lets carbon dioxide pass to the palisade cells and the oxygen is then able to pass freely back out through the use of air spaces. ✓ ⑪ There are also veins present at this part of the leaf which is where the transport of molecules around the plant takes place. The xylem cells bring water to the leaf and the phloem cells take the products of photosynthesis to the non-photosynthetic parts of the plant such as the roots. ✓ ⑨ The lower epidermis does not have any chloroplasts as it does not receive light but in between the epidermis cells are pairs of guard cells which make stomata. These are able to exchange gas and water. ✓ ⑫

Seren's answer

The leaf is adapted for gas exchange and photosynthesis in many ways. The leaf is thin ✓ ⑩ which provides short diffusion paths. Its waxy cuticle on the upper surface is transparent so light can penetrate ✓ ④ the palisade mesophyll layer. This layer is packed with chloroplasts ✓ ⑥ to absorb maximum sunlight. These chloroplasts can move and rotate ✓ ⑦ in order to absorb as much sunlight as possible. The spongy mesophyll layer has intercellular spaces between cells which allow the circulation of carbon dioxide. ✓ ⑪ These cells also contain some chloroplasts but not as much as the palisade cells which are elongated. The stomatal pores on the lower surface of the leaf permit the movement of gases ✓ ⑫ in and out of the leaf. The pores are surrounded by guard cells which close in the night and open in the day. ✗ ⑭ The leaf also has a large surface area to absorb ✓ ① as much sunlight as possible. The leaf can also move slightly at an angle ✓ ② where its surface captures maximum sunlight. The waxy cuticle prevents water loss. ✗ ⑬

Examiner commentary

⑥ Tom correctly states that the palisade mesophyll cells are packed with chloroplasts but fails to say why this is an adaptation.

⑧ He does not state how the spongy mesophyll cells are well suited for their function.

Summative comment

Tom's answer lacks planning and detail. He also makes a number of vague points. If he had better expression and expanded on some of his points he could have achieved more marks.

Tom achieves 5 out 10 marks.

Examiner commentary

⑭ Seren states that the stomatal pores close at night but fails to state that this reduces water loss.

⑬ The waxy cuticle does not prevent water loss, it reduces it.

Summative comment

Seren's answer lacks organisation, nevertheless she has made a number of good points.

Seren achieves 8 out of 10 marks.

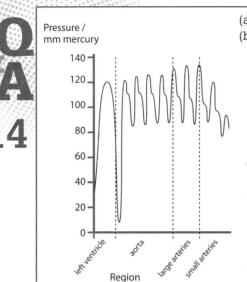

Q&A 14

(a) Describe the function of capillaries. *(3 marks)*
(b) The diagram shows pressure changes recorded as blood flows through the heart and the arteries.
  (i) What is the value of the systolic pressure in the left ventricle? *(1 mark)*
  (ii) Fluids flow from regions of high pressure to regions of low pressure. The minimum pressure in the ventricle is lower than the minimum pressure in the aorta. Explain why blood does not flow back into the ventricle from the aorta. *(2 marks)*
  (iii) Explain what causes the left ventricle pressure to fall to a very low value. *(2 marks)*
  (iv) If the diagram had been extended to include the flow through the capillaries in the body, give two ways in which the trace would differ from the diagram. *(2 marks)*
  (v) Give one reason why the pressure in veins is lower than in capillaries. *(1 mark)*
  (vi) How is flow maintained at this low venous pressure? *(1 mark)*

Tom's answer

(a) They are thin walled to let oxygen diffuse out to the cells. They also have a small diameter which slows down the flow of blood. ✓ ✓ ①
(b) (i) 120 ✗ ②
   (ii) Because of valves ✗ ③
   (iii) The ventricle relaxes ✗ ④
   (iv) The line would go down to a much lower level ✓ ✗ ⑤
   (v) They have a much bigger diameter ✓
   (vi) The muscles squeeze the blood up ✓

Examiner commentary

Tom's responses lack detail and are poorly expressed. If he expanded his answers he could achieve higher marks.

① Tom gains two marks out of the three as he has not stated that the walls are permeable.

② Tom has failed to include the units and so gains no marks.

③ He does not name the valves or their position in the heart.

④ If Tom had expanded his answer to state that the relaxation of the ventricle increases the volume and that the volume increases more rapidly than it is being refilled, he would have gained the mark.

⑤ Tom has given only one reason and gains one mark.

Tom achieves 5 out of 12 marks.

Seren's answer

(a) Capillaries are thin-walled consisting of a single layer of epithelial cells. Their walls are permeable so that exchange of materials between the blood and tissues takes place by diffusion. They also have a small diameter which slows blood flow allowing time for diffusion of oxygen and glucose to take place. ✓ ✓ ✓
(b) (i) 120mmHg ✓
   (ii) There are semilunar valves at the beginning of the aorta. These close under aortic pressure preventing backflow. ✓ ✓
   (iii) The ventricle relaxes, increasing its volume. As the volume increases more rapidly than it is being refilled, the pressure decreases. ✓ ✓
   (iv) The pressure line on the graph would go down to a lower level and there would be no rhythmical fluctuation. ✓ ✓
   (v) Veins have a much larger diameter. ✓
   (vi) The massaging effect of muscles and there are valves to prevent backflow. ✓

Examiner commentary

Seren has given detailed answers showing a good understanding of blood flow through the heart.

Seren achieves 12 out of 12 marks.

Q & A

15

(a) Explain how two features of a red blood cell (erythrocyte) enable it to carry out its function. *(2 marks)*

The graph below shows the oxygen dissociation curve for normal adult human haemoglobin (A) and *Arenicola* (lugworm) haemoglobin (B). The lugworm lives in muddy sand on the seashore.

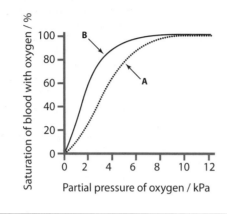

(b) What is the advantage of the S-shaped curve shown by haemoglobin? *(2 marks)*
   (i) In the tissues?
   (ii) In the lungs?

(c) (i) Describe the effect on the human haemoglobin line (A) if the carbon dioxide concentration was increased. *(1 mark)*
   (ii) Name this effect. *(1 mark)*

(d) (i) Lugworm has a curve to the left of human haemoglobin. What is the advantage of this to the lugworm? *(1 mark)*
   (ii) What does this suggest about the conditions under which the lugworm lives? *(1 mark)*

Tom's answer

(a) The red blood cell has a large surface area to carry oxygen. ⊗

It has no nucleus so has space for carrying more oxygen. ⊗ ①

(b)(i) More oxygen is released ✔ ⊗ ②

   (ii) It is saturated at low partial pressures. ⊗ ③

(c)(i) It moves to the right. ✔

   (ii) Chloride shift. Bohr Effect. ⊗ ④

(d)(i) It is quicker at gaining oxygen. ⑤

   (ii) There is less oxygen as it gets waterlogged. ✔

Examiner commentary

① Tom's answer is incorrect as the increased surface area is for the uptake of oxygen. His second point gains no marks as the extra space is to contain haemoglobin (which in turn combines with oxygen).

② Tom gains one mark but fails to gain the second mark as he does not explain why more oxygen is released.

③ Tom is careless and fails to identify what is saturated.

④ Tom is unsure and has given two answers. His first answer only is marked and is incorrect.

⑤ It is incorrect to state that haemoglobin is quicker at combining with oxygen.

Tom achieves 3 out of 8 marks.

Seren's answer

(a) The red blood cell is shaped like a biconcave disc to increase the surface area for the uptake of oxygen. ✔

It has no nucleus so has more space for haemoglobin. ✔

(b) (i) As the partial pressure of oxygen decreases more oxygen is released. ✔

   (ii) Haemoglobin is saturated at low partial pressures. ✔

(c) (i) The line moves to the right of line A. ✔

   (ii) Bohr effect ✔

(d) (i) Its haemoglobin has a greater affinity for oxygen. ✔

   (ii) It lives under conditions of low levels of oxygen. ✔

Examiner commentary

Seren has an excellent understanding of this difficult topic. In (a) alternative responses are that a RBC has a flexible shape allowing the cells to squeeze through capillaries / contains haemoglobin which combines readily with oxygen.

Seren makes correct use of terms such as 'saturation' and 'affinity'. Where carbon dioxide levels are high, as in muscle tissue, the curve shifts to the right, so haemoglobin has a greater affinity for carbon dioxide than for oxygen, and so oxygen is released to the muscle. She understands that in conditions where the partial pressure of oxygen is particularly low, as in the case of the lugworm, the curve is situated more to the left than that in the human, enabling it to load oxygen more readily.

Seren achieves 8 out of 8 marks.

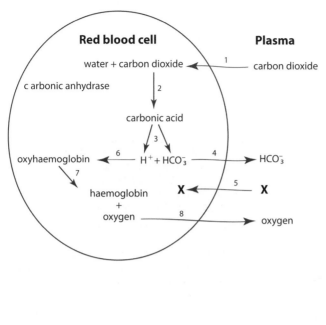

Red blood cell   **Plasma**

water + carbon dioxide ← 1 — carbon dioxide

c arbonic anhydrase

2

carbonic acid

3

oxyhaemoglobin ← 6 — $H^+ + HCO_3^-$ — 4 → HCO_3^-

7

haemoglobin **X** ← 5 — **X**
+
oxygen — 8 → oxygen

(a) State the main way that carbon dioxide is transported in the plasma. *(1 mark)*

(b) The diagram shows the role played by red blood cells in the transport of carbon dioxide. The red blood cell shown is located in respiring tissue.

(i) Carbon dioxide dissolves only very slowly in water. Inside red blood cells it dissolves quickly. Use information from the diagram to explain why carbon dioxide dissolves more quickly inside red blood cells. *(1 mark)*

(ii) Name ions X which enter the red blood cells. *(1 mark)*

(iii) Explain why ions X move into the red blood cells. *(1 mark)*

(iv) Use steps 6, 7 and 8 to explain why a high carbon dioxide concentration in tissues causes more oxygen to be released by red blood cells. *(2 marks)*

Tom's answer

(a) As a solution ⊗ ①

(b) (i) Because of the enzyme ✔

(ii) Chlorine ⊗ ②

(iii) To neutralise the pH ⊗ ③

(iv) Hydrogen ion joins with oxyhaemoglobin to give oxyhaemoglobin and releasing oxygen. ⊗ ④

Examiner commentary

Tom has a poor overall understanding of this process.

① Carbon dioxide is transported in three main ways, the main one being hydrogen carbonate.

② Chlorine fails to gain a mark, he should state chloride ions

③ Tom has become confused with the dissociation of carbonic acid into hydrogen ions and hydrogen carbonate and thinks that the overall pH of the cell is affected. In fact as hydrogen carbonate ions move out, chloride ions move in to maintain electrochemical neutrality.

④ The release of hydrogen ions creates more acidic conditions in the red blood cell which causes oxyhaemoglobin to dissociate releasing oxygen with the formation of HHb. The oxygen diffuses out to the tissues.

Tom achieves 1 out of 6 marks.

Seren's answer

(a) As sodium hydrogen carbonate ✔

(b) (i) Because of the enzyme, carbonic anhydrase ✔

(ii) Chloride ions ✔

(iii) To counteract the negatively charged hydrogen carbonate ions that have moved out. ✔

(iv) Carbonic acid dissociates into hydrogen ions and bicarbonate ions. The acid condition causes the release of oxygen. ⊗ ⊗ ①

Examiner commentary

① Seren has not used the information provided in the diagram to explain that hydrogen ions provide the conditions for the oxyhaemoglobin to dissociate into oxygen and haemoglobin. The hydrogen ions combine with haemoglobin to form haemoglobinic acid. The dissociated oxygen is released from the red blood cell to the tissues.

Seren achieves 4 out of 6 marks.

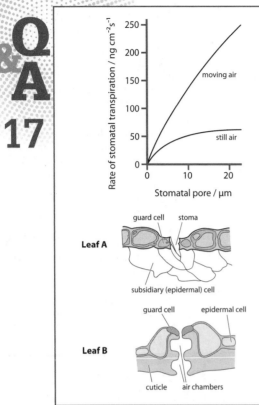

Q&A 17

(a) The graph shows the relationship between the rate of transpiration and the diameter of the stomatal pores in still and moving air.

(i) Describe the relationship between the rate of transpiration and the stomatal diameter in still air. *(3 marks)*

(ii) Explain the differences between transpiration rates in still and moving air. *(3 marks)*

(b) Each diagram below shows a section through the epidermis of a leaf. Leaf A is from a plant that grows in areas of regular rainfall. Leaf B is from a plant that grows in very dry conditions.

(i) Describe two features of leaf B that enable the plant to live in very dry conditions. *(2 marks)*

(ii) Explain how each of these features is effective. *(2 marks)*

(iii) Describe and explain one other feature, not shown in leaf B, which may be present in such plants to reduce the amount of water vapour lost by transpiration. *(2 marks)*

(c) (i) State the general name given to plants that are adapted to living in dry conditions. *(1 mark)*

(ii) State two environmental factors that can decrease the rate of transpiration in a plant. *(2 marks)*

Tom's answer

(a) (i) As the transpiration rate increases, stomatal pore gets bigger but it levels off.
Ⓧ ①

(ii) In moving air, moisture is blown away from inside the leaves through the stomata.
Ⓧ ②

(b) (i) Thick cuticle and sunken stomata.
✓ ✓ ③

(ii) The cuticle makes a waterproof layer.
Ⓧ ④

Water is trapped in the air chambers above the stomata so no wind can get to it.
Ⓧ ⑤

(iii) Rolled leaf to trap moisture.
✓ Ⓧ ⑥

(c) (i) Zerophyte
✓ ⑦

(ii) Temperature and humidity
Ⓧ Ⓧ ⑧

Examiner commentary

Although Tom has attempted all the questions, his answers are vague and lack the detail required at this level. He has taken little notice of the mark allocation to each sub question. For example, in (a) (i) and (ii) there are three marks allocated to each so he should make three points. He has made only one point in each case.

① Tom gains no marks as he has misread the graph. He makes no reference to data from the graph and does not refer to the levelling off or plateau.

② Tom incorrectly refers to 'moisture' rather than 'water vapour'. His answer is far too vague and there is no reference to there being no increase in the rate after a certain pore size.

③ He has failed to make a comparison with leaf A that the cuticle is thicker but he has gained the mark.

④ The cuticle is not a waterproof layer, it does not prevent water loss, it reduces water loss.

⑤ Tom has not referred to water vapour and so does not gain the mark.

⑥ Tom has made an improved reference to water being moisture, but this is not acceptable at this level.

⑦ Although Tom has misspelt xerophyte the mark is given as the spelling is phonetic.

⑧ Tom has failed to read the question correctly. The question does not merely ask for factors that affect the rate but a description of the factors that *decrease* the rate.

Tom has achieved 4 out of 15 marks.

Seren's answer

(a) (i) As the diameter of the stomatal pore increases, the transpiration rate increases. ✓ But at a certain diameter the rate starts to slow down and levels off. ✓ A point is reached where there is no increase as the diameter increases. ✓

(ii) The air inside the leaf is saturated with water vapour. ✓ In still air water vapour evaporates through the stomata but forms a boundary layer over the pore. There exists equilibrium between the water potential inside the leaf and the atmosphere. ✓ When it is windy the water vapour is blown away and a water potential gradient is set up between the ✓ air spaces in the leaf and the surrounding air. As a result the rate of transpiration continues to increase.

(b) (i) Thicker cuticle and sunken stomata within a deep air chamber. ✓ ✓

(ii) The thick cuticle reduces the loss of water through the epidermis. ✓ Water vapour is trapped above the stoma and reduces the water potential gradient. ✓

(iii) Hairs above the stoma to trap water vapour. ✓ ✓

(c) (i) Xerophyte ✓

(ii) Low temperature and a decrease in wind speed. ✓ ✓

Examiner commentary

In (a) (i) and (ii) Se marks. She could h answer by using fig graph to illustrate h

Seren uses the corre ...ater vapour' and uses her knowledge of water potential correctly to introduce the idea of a gradient between inside the leaf and the atmosphere. She refers to a 'boundary layer' over the pore. An alternative term is 'diffusion shells'.

In (b) her answers are concise and detailed.

Seren achieves 15 out of 15 marks.

Q&A

18

Reproduction

(a) Many organisms can reproduce sexually and asexually. Give one advantage and one disadvantage of sexual reproduction. *(2 marks)*

(b) Explain what is meant by the term 'internal fertilisation'. *(1 mark)*

(c) Give three advantages of internal fertilisation and development to terrestrial animals. *(3 marks)*

(d) Suggest two reasons why flowering plants have been so successful in the colonisation of land. *(2 marks)*

Tom's answer

(a) Sexual reproduction involves two parents and it produces young that are genetically different. ✗ ①

(b) The sperm is placed inside the female. ✗ ②

(c) There is less chance of the sperm being wasted and there is less chance of them drying out. Also the baby is protected inside the mother. ✓ ③

(d) They can reproduce sexually and asexually. ✗ ④

Examiner commentary

① Tom has given two valid answers but has failed to state whether these are an advantage or a disadvantage.

② The answer is far too vague at this level. He makes no reference to the process of fertilisation, i.e. the fusion of the sperm and egg.

③ He gains one mark for stating that there is less chance of sperm being wasted. However, he has not fully answered the question regarding the reference to 'terrestrial animals' in that gametes are independent of water. His third point referring to mother and baby does not warrant a mark at this level.

④ Tom cannot think of the correct answer and has resorted to repeating the stem of the question.

Tom achieves 1 out of 8 marks.

s answer

An advantage of sexual reproduction is that it involves two parents and this introduces variation because gametes are produced by meiosis and so the offspring are genetically different. A disadvantage is that mating has to take place and this can be a slow form of reproduction. ✓ ✓

(b) This is the fusion of gametes, where the sperm fertilises the egg inside the female's body. ✓

(c) There is an increased chance of fertilisation and fewer sperm are wasted than would be in the water. The sperm do not have to rely on water in which to swim so they are not likely to dehydrate. The fertilised egg or zygote develops inside the body of the female where it is nourished and protected. ✓ ✓ ✓

(d) The life cycle is quite rapid and their seed has a food store so that the young plant or embryo has a good start. The seed also has a resistant outer layer or seed coat which enables it to survive unfavourable periods, such as the winter. ✓ ✓

Examiner commentary

Seren's responses are concise yet detailed. She has made connections with other topics such as meiosis. In (d) she could have referred to the link between flowering plants and insects. Also, that when leaf fall occurs and they decay nutrients are recycled into the soil. A large number of seeds are also produced.

Seren achieves 8 out of 8 marks.

Q & A 19

(a) Compare the dentition of a carnivore and a grazing herbivore.
(b) Describe how the gut region of a ruminant is adapted to its diet.

(10 marks)

Mark scheme

A candidate would be expected to give any 10 of the following 15 marking points.

① Herbivore – small, flat top incisors / horny pad ② canines absent ③ teeth continuously growing ④ cheek teeth, large surface area for grinding / WM arrangement ⑤ diastema qualified ⑥ carnivore – canines large ⑦ carnassials qualified ⑧ sharp cutting edges on teeth ⑨ herbivore – side to side jaw movement / carnivore – up and down jaw movement ⑩ four chambers ⑪ large volume / longer gut ⑫ symbiotic bacteria / mutualism ⑬ anaerobic conditions ⑭ regurgitation of food / chewing cud qualified ⑮ digestion of cellulose / digestion by bacteria

Tom's answer

(a) A carnivore has carnassial teeth. ✓ ⑦ They slide past each other instead of coming together. Carnivores also have very large canine teeth to grab onto their prey and stop them from moving. ✓ ⑥ The molars of a carnivore are incredibly sharp compared to normal molars. A grazing herbivore has molar teeth that are used to grind vegetation to increase its surface area. The jaw moves in a side-to-side motion ✓ ⑨ instead of up and down, which helps grind the vegetation. Most of the grazing herbivore's teeth are molars or premolars and are large and flat to help grind their food. ✗ ④ You would see an example of this dentition in sheep or cows.

(b) The diet of a ruminant is vegetation, i.e. grass that is made of cellulose cell walls which are very hard to break down and digest. ✓ ⑮ Because of this, a ruminant has adapted to its diet by first chewing and grinding the vegetation to increase the surface area of the food and here salivary amylase ✗ ⑫ is added to start breaking down the cellulose. The bolus is then swallowed and enters the rumen of the cow's stomach where the chemical and mechanical digestion takes place as the food is moved from the rumen and into the reticulum. This increases the surface area further and the addition of acid in the stomach breaks down the vegetation further. The vegetation is then regurgitated and then re-chewed by the ruminant. ✓ ⑭ This adds more salivary amylase and grinds the food more into cud. The cud is then re-swallowed and goes into the ornasum of the ruminant's stomach where most of the water is absorbed and a lot of nutrients are absorbed.

Examiner commentary

④ Tom refers to the cheek teeth of the herbivore as being large and flat but makes no reference to the increased surface area of the teeth for grinding the food. He is confused on this point as earlier he mentions that the teeth increase the surface area of the food.

⑦ He refers to carnassials in the carnivore but does not qualify this with their function or the way they operate.

⑫ Tom incorrectly states that the enzyme, salivary amylase breaks down cellulose, rather than cellulase.

Summative comment

Tom has made a number of vague and confused statements. He obviously has ability but has failed to learn his work in sufficient detail.

Tom achieves 5 out of 10 marks.

Seren's answer

(a) Carnivores such as tigers have large specialised cheek teeth called carnassials which slide past each other. ✔ ⑦ They have sharp incisors to grip and tear flesh from bone, and large canines. ✔ ⑥ They have a powerful jaw which moves up and down. Their dentition is adapted to rip meat and catch prey. A grazing herbivore has very different dentition. It has no carnassial teeth and no need for sharp teeth, as it does not eat meat. It has no incisors on its top jaw only a hard plate. ✔ ① A herbivore also has a jaw which moves from side to side ✔ ⑨ to allow grinding of vegetation using its flat molars. It also has a gap called a diastema in its teeth to allow the tongue room to separate newly chewed food from cud. ✔ ⑤

(b) An example of a herbivore is a cow, which is also a ruminant. A ruminant has a gut which is specially adapted for grazing and digesting plant material. The four regions of the gut include the rumen. Ruminants do not have the ability to produce the enzyme cellulase which breaks down cellulose ✔ ⑮ (in the cell walls of plants), so in a region of the gut it has a special relationship, called commensalism, ✔ ⑫ with bacteria which produce this enzyme for it. Food is regurgitated from the second region and chewed again. ✔ ⑭ This is called the cud. It is then swallowed and passes to the third region where water is absorbed. The fourth region of a ruminant gut acts as a normal stomach.

Summative comment

Seren makes some attempt at comparing the dentition of a carnivore and a herbivore. She understands the symbiotic relationship between the ruminant and the bacteria and their importance in the digestion of cellulose but does not answer the question regarding the adaptations of the gut itself.

Seren achieves 8 out of 10 marks.

Q&A 20

(a) Define the term parasite. *(2 marks)*

(b) Name two characteristics of tapeworms which are adaptations to their parasitic life. *(2 marks)*

(c) The tapeworm does not possess an alimentary canal (gut). How does the animal survive without this system? *(2 marks)*

(d) (i) Name two features of the tapeworm's reproductive system which are adaptations to its parasitic existence. *(2 marks)*

(ii) Explain the importance of each of these features in the tapeworm. *(2 marks)*

Tom's answer

(a) A parasite is an organism that lives inside another organism. ✔ ①

(b) It has suckers and a thick outer coat. ✔ ✗ ②

(c) It takes in materials through its surface. ✗ ③

(d) (i) It produces lots of eggs and embryos. ✔ ④

(ii) Many eggs are wasted as they do not get fertilised. ✗ ⑤

Examiner commentary

① Tom has not given a full definition and has gained one of the two marks available.

② He gains one mark for the term 'suckers' but he has not used the term 'cuticle' for the second mark.

③ His answer contains no detail.

④ Eggs and embryos are the same marking point.

⑤ Again Tom has not expanded on the point he has made and has failed to mention that the tapeworm has two hosts.

Tom achieves 3 out of 10 marks.

Seren's answer

(a) A parasite is an organism that lives in or on another organism, called its host, and gains nourishment from the host and usually causes it harm. ✔ ✔

(b) It is hooks or suckers and a thick cuticle. ✔ ✔

(c) It is bathed in digestive juices of the host and absorbs nutrients directly through the body surface from the gut. ✔ ✔

(d) (i) It produces large numbers of eggs and embryos. The tapeworm is so big the gut could not contain more than one individual so each segment contains both male and female sex organs. ✔ ✔

(ii) The tapeworm has a secondary host, the pig, so it has to produce many eggs and embryos as there is high mortality when the offspring are transferred from one host to another. As mating between two individuals cannot take place the tapeworm fertilises its own eggs. ✔ ✔

Examiner commentary

Seren has given concise, detailed responses. An alternative answer to (b) in this context is that the tapeworm has inhibitors on its surface to prevent digestion by the host's enzymes.

Seren has achieved 10 out of 10 marks.

Quickfire answers

BY1 Basic Biochemistry and Cell Structure

① Condensation – when two monosaccharides join to form a disaccharide with the elimination of water and the formation of a glycosidic bond.
Hydrolysis – the addition of water to a disaccharide resulting in the formation of two monosaccharides.

② Lactose – disaccharide – plants and animals – energy
Cellulose – polysaccharide – plants – structural
Glucose – monosaccharide – plants and animals – energy
Glycogen – polysaccharide – animals – energy store

③ Broken down to form one molecule of glycerol and three molecules of fatty acid with the elimination of water and the breaking of an ester bond.

④ When oxidised, lipids provide more than twice as much energy as carbohydrates. If fat is stored, the same amount of energy can be provided for less than half the mass. It is therefore a lighter storage product.

⑤ Water produced from the oxidation of food is called metabolic water. When water is scarce a camel can metabolise fats.

⑥ A triglyceride has three fatty acids, no phosphate group and is non-polar.
A phospholipid has two fatty acids, one phosphate group and a hydrophilic head and a hydrophobic tail.

⑦ Hydrophilic heads.

⑧ 1. Intrinsic protein
2. Phospholipid

⑨ Peptide
Hydrogen
Disulphide
Ionic

⑩ Insulin – globular
Collagen – fibrous
Keratin – fibrous
Lysozyme – globular

⑪ A polar molecule carries an unequal distribution of electrical charge. The oxygen end has a slightly negative charge and the hydrogen end of the molecule has a slightly positive charge.

⑫ Pondskater – high surface tension.
Water has a maximum density at 4°C so ice floats forming an insulating layer for animals below.
Sweating – high latent heat.

⑬ 1. To allow chemical reactions to take place in solution.
2. To allow light to pass through for photosynthesis.

⑭ Core of nucleic acid
Protein coat

⑮ 1. Ribosome
2. Golgi body
3. Mitochondrion
4. Nucleolus

⑯ 1. Mitochondria
2. Chloroplast
3. Endoplasmic reticulum

⑰ Chloroplasts
Cell wall
Large permanent vacuole

⑱ Kidney – organ
Epithelium – tissue
Muscle – tissue
Sperm – cell

⑲ Extrinsic proteins occur on the surface of the bilayer or partly embedded in it.
Intrinsic proteins extend across both layers.

⑳ Vitamin A is fat soluble and freely passes through the lipid bilayer. Large molecules such as glucose are insoluble in lipids and cannot pass through the non-polar centre of the phospholipid bilayer. Intrinsic proteins assist glucose to pass in and out of the cell by facilitated diffusion.

㉑ Larger surface area; thin.

㉒ Increase in kinetic energy of molecules results in an increase in rate.

㉓ Carrier proteins are involved in facilitated diffusion.

㉔ Both use carrier proteins; active transport requires energy/ATP; occurs against a concentration gradient.

㉕ Muscle contraction
Nerve impulse transmission
Protein synthesis
Uptake of minerals by roots

㉖ CAB.

㉗ When the cell membrane just pulls away from the cell wall.

㉘ A rise in temperature increases the kinetic energy of molecules. In an enzyme catalysed reaction the enzyme and substrate molecules collide more often in a given time so that the rate of reaction increases.

㉙ When all the active sites are filled, a point is reached when all the active sites are working as fast as possible. The rate of reaction is at a maximum and the addition of more substrate will have no effect on the rate of reaction, which levels off.

㉚ pH
Substrate concentration
(Not temperature)

㉛ In competitive inhibition the inhibitor occupies the active site of the enzyme. In non-competitive inhibition the inhibitor attaches to a site other than the active site.

㉜ Increasing the substrate concentration reduces the effect of the competitor inhibitor.
In non-competitive inhibition the rate of reaction is unaffected.

㉝ The rate of reaction of the immobilised enzyme is greater between 0°C and 40°C; the optimum temperature of the immobilised enzyme covers a wider range; the immobilised enzyme begins to denature at a higher temperature/the free enzyme is completely denatured at 60°C whereas the immobilised enzyme is denatured at 80°C; the immobilised enzyme is more active at all temperatures except 40°C.

BY2 Biodiversity and Physiology of Body Systems

㉞ Accuracy
Quantitative result
Measures low concentrations

㉟ Cytoplasm only – ribosomal RNA and transfer RNA
Nucleus and cytoplasm – mRNA

㊱ DNA – sugar is deoxyribose, double stranded, thymine base
RNA sugar is ribose, single stranded, uracil base

㊲ Replication of DNA
Cell increases in size
Organelles produced replacing those lost during previous division
ATP production

㊳ Production of large numbers of identical offspring in a relatively short period of time, e.g. bulbs, tubers and runners.

㊴ Mitosis – one division, chromosome number unchanged, no crossing over, daughter cells genetically identical.
Meiosis – two divisions, chromosome number halved, crossing over occurs, daughter cells genetically different.

㊵ Crossing over
Independent assortment
Mixing of two parental genotypes

① Any two of:
Loss of habitat
Over-hunting by humans
Competition from introduced species
Deforestation

② A group of organisms which can interbreed to produce fertile offspring.

③ The binomial system uses an international language so that scientists are presented with precise identification worldwide.

④ They do not possess chlorophyll.

⑤ Any two of:
Support
Reduced water loss
Protection

⑥ Hippo and cows

⑦ *Macroderma gigas*

⑧ Homologous

⑨ Surface area
Thickness of membrane
Permeability
Concentration gradient
Temperature

⑩ With increase in size there is a higher metabolic rate and a greater requirement for oxygen for respiration. There is also a smaller surface area to volume ratio so oxygen needs to diffuse over a greater distance.

⑪ Gills
Lungs
Tracheae

⑫ Large surface area
Thin
Permeable
Good blood supply

⑬ Maintains a concentration gradient over the whole length of the gill filament. Also, exchange occurs over a longer period.

⑭ By having valves that can close spiracles when the insect is inactive.

⑮ Because it relies on diffusion to bring oxygen to the respiring tissues. If insects were larger, it would take too long for oxygen to reach the tissues sufficiently rapidly to supply their needs.

⑯ Rapid
No respiratory pigment required
Reduced water loss
Oxygen supplied directly to tissues
No transport system required

⑰ Alveoli
Bronchioles
Bronchus
Trachea

⑱ Preventing trachea collapsing during inspiration.

⑲ Palisade mesophyll – traps sunlight/ site of photosynthesis.
Spongy mesophyll – exchange of gases from atmosphere to chloroplasts of palisade mesophyll.

⑳ Insects may create mass flow, plants do not.
Insects have a smaller surface area to volume ratio than plants.
Insects have tracheae along which gases can diffuse, plants do not.
Insects do not interchange gases between respiration and photosynthesis whereas plants do.

㉑ Chloroplasts and uneven thickening of inner and outer walls.

㉒ Medium, e.g. blood
Pump, e.g. heart
Respiratory pigment, e.g. haemoglobin

㉓ No pigment in insect, haemoglobin in mammals
Open blood system in insects, closed in mammals
O_2 is transported through trachea, not via blood

㉔ In passing through the heart the second time a double circulation ensures that the blood is pumped around the body at a greater pressure.

㉕ To withstand the pressure when blood is pumped from the left ventricle of the heart

To maintain the pressure, which can be reduced by friction when blood passes through the vessels

㉖ To slow blood flow and allow time for diffusion of substances.

㉗ Greater force of blood from the base upwards and also ensures that the compartment is completely emptied.

㉘ Haemoglobin

Large surface area – biconcave shape

No nucleus – more room to contain haemoglobin

㉙ A theoretical line would be proportional bisecting the axis whereas the actual line shows a steeper rise between 2Kpa and 7Kpa.

The theoretical line does not flatten at the top.

㉚ It is more to the left; it combines more readily with oxygen. It has a higher affinity with oxygen and so takes up oxygen from the mother's haemoglobin.

㉛ Increase in the number of red blood cells.

㉜ To preserve electrochemical neutrality

㉝ Hydrostatic pressure and water potential/ osmosis

㉞ Venous end and lymphatic system

㉟ Support and transport of water

㊱ Soil solution has a high water potential, the root hair cells have a lower water potential and water moves in by osmosis.

㊲ Endodermal cells possess a Casparian strip or waterproof band that prevents the passage of water.

㊳ Reduced or stopped as it is an energy-requiring process.

㊴ 1. Increase

2. Increase

3. Decrease

㊵ Water evaporation from the leaves is trapped and the region above the stomata aperture becomes saturated with water vapour. The water potential gradient between the inside of the leaf and the outside is reduced.

㊶ Cuticle

Ability to close gas exchange system openings: spiracles in insects, stomata in plants

㊷ Sieve tubes and companion cells

㊸ To provide energy for the sieve tubes as they have lost their cell organelles.

㊹ *Advantages*:

Variation/genetically different

Allows the development of a resistant stage in the life cycle, e.g. seeds, spores, larvae.

Disadvantages:

Need two individuals

Slow

Mutations or genetic disorders

Need for gametes to fuse

Large numbers produced

㊺ Reduced number of gametes

More chance of fertilisation; less chance of gametes wasted

Male gamete becomes independent of water

Embryo and zygote can be better protected

Resistant stage in life cycle

㊻ Egg with shell

Internal fertilisation

㊼ Reproduction using pollen

Efficient water-carrying xylem vessels

Food store in seeds allows rapid growth in embryo/enables seed to survive for long periods

No need for water for fertilisation

Leaves fall and decay for recycling of nutrients

㊽ Fungi and bacteria

㊾ The breakdown of large insoluble molecules into soluble molecules by means of enzymes.

㊿ Serosa

Muscle

Submucosa

Mucosa

51 Carbohydrate – glucose

Fats – fatty acids and glycerol

Proteins – amino acids

52 1. Bile emulsifies fats, i.e. breaks down fat into smaller droplets having a larger surface area for more efficient enzyme action. Also neutralises acid from the stomach.

2. Mucus lubrication and protection.

53 Glucose and amino acids – capillaries

Fatty acids and glycerol – lacteal

54 Sheep – molars and premolars with large cusps for grinding/no carnassials/large incisors/no canines.

Dog – carnassials/shape of incisor/large canines

55 Cows lack the ability to produce cellulase and since plants contain cellulose cell walls, they rely on bacteria to break down the cellulose.

56 pH extremes

Immune system of host

Digestive juices

Constant movement of digestive juices

57 Hooks/suckers

Thick cuticle